BRINGING AERODYNAMICS TO AMERICA

BRINGING AERODYNAMICS TO AMERICA

Paul A. Hanle

The MIT Press
Cambridge, Massachusetts
London, England

This book was set in Janson by Achorn Graphic Services Inc., and
printed and bound in the United States of America.

Library of Congress Cataloging in Publication Data

Hanle, Paul A.
  Bringing aerodynamics to America.

  Includes bibliographical references and index.
  1. Aeronautical research—United States—History.
2. Aerodynamics—History. 3. Von Kármán, Theodore, 1881–1963.
I. Title
TL565.H32   629.132′3′09   81-20690
ISBN 0-262-08114-8   AACR2

The Theodore von Kármán Collection at the California Institute of
Technology Archives has been copied on microfiche through a contract
from the National Air and Space Museum. Scholars may consult a copy
of this microfiche edition, which is deposited in the National Air and
Space Museum library as well as at Caltech.

CONTENTS

PREFACE

This book is about aerodynamicist Theodore von Kármán's move from Germany to America in the 1920s. That move, though of interest as a part of the biography of a singular aviation pioneer, has occupied my attention for another reason. I have taken it as the focal point for discussing the origins of American expertise in theoretical aerodynamics, an expertise that has contributed to the success and strength of the American aviation industry. These origins are my main concern, for they reveal the "scientific" part of aviation history as a cultural phenomenon within the larger historical development of science-based technology in America in this century.

Kármán not only set the course of research at the Daniel Guggenheim Graduate School of Aeronautics at the California Institute of Technology after its founding in 1926, but he brought with him a kind of academic entrepreneurship of aviation—pursuing basic scientific research in aerodynamics while teaching, giving popular talks, and helping to build airplanes—that was characteristic of applied science in the European setting from which he had emerged. In bringing Theodore von Kármán to America, the leaders of Caltech brought the organization and personality of Göttingen-style applied science.

Caltech's aeronautics grew and flourished in the fifteen years after Kármán's first visits. If one cannot trace all of theoretical aerodynamic activities in the United States directly to Kármán and his group by 1940 (one can trace *nearly* all), it is clear that the group was peerless among its few American rivals in international influence. Kármán's influence grew among the military and brought lucrative consulting for himself and his colleagues. Airplanes, notably the DC series, were tested in his wind tunnel, and some were even designed at the school. Caltech produced few working aeronautical engineers, but its students dominated academic aerodynamics. Kármán and his school became the national leader of "scientific" aeronautics, and by himself Kármán led theoretical aerodynamics.[1]*

Lest the reader protest my neglect of the influence of aerodynamics pioneers like Jerome C. Hunsaker of the Massachusetts Institute of Technology (MIT), William F. Durand of Stanford University, Max Munk of the U.S. National Advisory Committee for Aeronautics (NACA), and even the venerable Orville Wright, I should specify the sense in which theoretical aerodynamics was the near exclusive haunt of Theodore von Kármán. The term *aerodynamics* had been applied in history with varying degrees of precision but when used most carefully it designated the study of the flow of air around obstacles. All studies of this flow were based on experiment, which of course required some connection to reasoning, to what physicists call semiempirical theory. Measurements of lift-to-drag ratios for different angles of attack of a wing, for example, could generate data for a semiempirical theory of lift with appropriate curves. What was almost completely absent in American aeronautical studies in 1926 was not this kind of theory but the sophisticated mathematical apparatus with which to build a theoretical function that corresponded to measurement. Such an apparatus had existed in Europe for decades, developed by Helmholtz, Rayleigh, Jeans, Poisson, and later by Kutta, Joukowski, and Prandtl, among others. Max Munk enjoyed several important successes along

---

*Notes will be found at the end of the book.

theoretical lines following his German countrymen, but the failure of his career at NACA and other factors reduced the import of his work. The roots of theoretical aerodynamics lay in German, French, and English physics. The point is not that American aerodynamicists could not understand the theory but that most did not *develop* it before 1926. After that time Kármán's laboratory became the mecca for theoretical studies in the United States.

Kármán's move to America was part of the overall expansion and improvement of American science that by different accounts had begun sometime about the end of the nineteenth century.[2] His story reveals one way in which the United States nurtured a large applied research community, a part of a greater American community of science and technology that was to become the dominant developer of science-based technology during and after World War II. If it is true, as I believe, that Kármán's was the seminal part in building the science of flight in America, then to document the conditions and events that brought Kármán here is to describe a major step in the professionalization of American aerodynamics and of aeronautics itself.

Note: The names of Kármán and Prandtl have often been misspelled in the original sources. When quoting material from these sources, I have consistently maintained the spellings of the original authors.

ACKNOWLEDGMENTS

I must first thank Otto Mayr of the National Museum of American History, who suggested that I study the history of a field such as aerodynamics that bridges science and engineering. Subsequently he encouraged my research and suggested scholars with whom I might confer. Foremost among these are Walter Vincenti and Nicholas Hoff at Stanford, Robert Kargon at Johns Hopkins, and Louis L. Bucciarelli at MIT, each of whom read the first draft of this work and commented in detail. Their remarks stimulated me to revise, expand, and I hope improve the work. Each I thank, but especially Professors Vincenti and Hoff, who helped to correct some failings of the manuscript regarding aerodynamics. Richard P. Hallion, my former colleague at the National Air and Space Museum, suggested that I examine the papers of the Daniel Guggenheim Fund in the Library of Congress, which proved most fruitful. He and Tom D. Crouch of the Aeronautics Department, National Air and Space Museum, also gave welcome bibliographic help. Paul Forman of the National Museum of American History, Joseph J. Corn of the Air and Space Museum and Stanford subsequently, and an anonymous reviewer pointed to particular lines of research, some of which I have followed. Judith Goodstein and Carolyn Harding, Institute Archivist and Assistant at Caltech in 1978, directed me to relevant documents and helped me to place them in the context of the growth of Caltech in the 1920s. I thank them all.

My research took me to several repositories of unpublished documents in Europe. I want to thank Bibliotheksoberrat Dr. Klaus Haenel and his assistant Frau Schukert at the Niedersächsische Staats- und Universitätsbibliothek in Göttingen for their advice and help in sorting through the Felix Klein papers. I owe a special debt of gratitude to Hr. Fritz Wilhelm Freyni at the Registrar's office of the Technical University of Munich who on three hours notice produced from storage registrarial files c. 1920 relating to Ludwig Prandtl's call to Munich. Archivists at the Bavarian State Archives, whom I also thank, had advised me of the possibility of their existence at the University. Frau Margret Nida-Rümelin, librarian of the Forschungsinstitut at the Deutsches Museum, steered me through the Library and Special Collections Department there. I thank her especially, and also the head of the manuscript division Dr. Heinrich.

In America, Alex Roland of the NASA History Office helped me identify relevant papers in the collection of the NACA at the National Archives. Those papers are now under the jurisdiction of the Polar and Scientific Archives, whose Archivist Sharon Gibbs provided copies of key documents, as always with a smile. I thank these and each of the librarians and archivists of the bibliographic repositories listed above. Lillian Kozloski typed the manuscript, in a time when she was otherwise preoccupied. I am most grateful to her.

I want to express my gratitude to S. Dillon Ripley, the Secretary of the Smithsonian Institution, for providing a grant from his Fluid Research Fund that enabled my first, crucial visit to the Caltech archives. A Smithsonian Research Award provided funds for my tour of German sources in 1979. At every turn the Smithsonian Institution has supported this work without hesitation. I consider myself most fortunate to be able to do research in this generous institution.

BRINGING AERODYNAMICS TO AMERICA

1

AVIATION, AERONAUTICAL ENGINEERING, AND FOUNDATIONS, 1926

The year 1926 marked a turning point for aircraft manufacturing in America. The end of World War I had precipitated the cancellation of contracts for thousands of military aircraft. In the words of aircraft-business analyst W. G. Cunningham, "November, 1918, with the Armistice, brought as complete a collapse as has ever been experienced by any industry. Within three days, more than $100,000,000 worth of orders was cancelled."[1] Aircraft production fell from more than 14,000 airplanes in 1918 to fewer than 300 in the low year of 1922; and even through 1925 the total remained well below 1000. But in 1926 the number jumped by 400 (with a boost of 300 for civil aircraft alone), and the next three years saw phenomenal increases to 2000, 4300, and 6200 during the so-called Lindbergh boom.[2]

Popular perception of the aviation industry was not always favorable. In 1919, for example, press criticism of wartime procurement led to investigation by a presidential commission of the "aircraft trust" and intimate relations between the manufacturers and government purchasing agents. The industry was not completely acquitted, and both government and industry publicized efforts to correct abuses.[3] Barnstorming by air exhibitionists not only frequently cost them their lives but painted a circus image of airplanes, hindering their commercial use. Airmail pilots of the U.S. Post lost their lives with notable frequency as well.

Within the industry prospects for growth remained unclear through 1924. The market so depended on military purchases of aircraft in 1921 that when the army and navy began building their own, the manufacturers' association strenuously objected and launched a powerful and successful lobbying campaign for reinstatement of contracts. Before the practice was rescinded, wrote one "historian" associated with the manufacturers, "Even those companies which were not on the verge of bankruptcy had about decided to retire."[4] An increase of military orders from 226 in 1922 to 687 the next year caused a small boom, but it dissipated as the number fell back to 317 in 1924.[5] In such a market, manufacturers did not dare expand, nor was there much incentive for building from new designs. In 1924 the National Advisory Committee for Aeronautics (NACA), a government organization strongly identified with the interests of the manufacturers, wrote, "The present American aircraft industry is but a shadow of that which existed at the time of the armistice."[6]

Within a year, in part because of such government-industry sympathies and because of the uproar caused by Army Air Service Colonel Billy Mitchell's accusations of incompetence in the use of aircraft for national defense, a second call arose to investigate the industry. The resulting recommendations of an investigative board headed by financier Dwight Morrow followed along solidly probusiness, Republican lines and proved pivotal for consolidation of the industry. Thus by 1926 aircraft manufacturers, though still feeling beleaguered, sensed imminent change, with a clearer prospect for growth of their industry.

Three government acts of tangible support, two following recommendations of the Morrow board, gave cause for the more sanguine view. In early 1925 Congress had passed the so-called Kelly act, which provided for contracts to aircraft operators to carry mail by air. In mid-1926 the Air Commerce Act established an Aeronautics Branch of the Commerce Department to regulate airlines and in general to promote civil air transportation. This meant the installation of aerial beacons, the establishment of radio networks, and, perhaps most important to the industry, the sanctioned monopolies of route assignments. Both the army and navy established five-year plans in Spring 1926, thus appearing to stabilize the market for military procurement and industrial production of aircraft.[7] In sum, then, 1925–26 offered the plane builders something they wanted but had failed to attain since the end of the war—a "comprehensive national air policy" of government support.[8] They recognized its significance long before it came into being, as their immediate response in increasing aircraft production proves.

The association by which these airplane builders were bound together was no monolithic cartel. Despite common interests they often competed, with strongly individualistic and regionalist outlooks. Donald Douglas, for example, embraced the idea that Southern California, with its "diligent labor" and fine weather, was the future center of aviation.[9] Remarks about New York as the hub of aviation emanated also from the strongholds of Sikorsky and Fairchild.[10]

Four main regions were being boosted as centers of aircraft manufacture around 1926, though the companies were growing generally westward in a movement that mirrored the receding of America's geographical frontier in the preceding century. In New York and New Jersey, the main builders were Vought of Long Island City, Sikorsky of College Point, Fairchild of Farmingdale, and Fokker's Atlantic Aircraft of Hasbrouck Heights, New Jersey. Consolidated Aircraft formed in 1923 and set up operations in Buffalo in 1924, while the Curtiss Company built its successful Condor transport in Hammondsport, New York. In the region broadly encompassed by the term *Midwest* were Wright–Dayton of Dayton, Ohio; Martin of Cleveland; Stinson, Stout, and Ford of Detroit (Stout and Ford merged in that year to mass produce metal trimotor airplanes); McDonnell of Milwaukee; and the many companies of Wichita, among them Cessna, Beech, and Stearman, that would form the core of the general-aviation industry. The West had Glenn Martin's works in Los Angeles and

Lockheed based in Burbank, but with his World Cruisers circling the globe in 1924 Santa Monica's Donald Douglas would soon dominate the Southern California market. Seattle, the center of the Pacific Northwest region, hosted William Boeing and his giant aircraft-airline conglomerate that would become United Aircraft and Transportation Corporation. The westward movement was evident in Consolidated, for example, which established a plant to build seaplanes on the shores of San Diego harbor.[11]

Aircraft design began anew after 1926, with the focus laid, as it had not been before, on building safe commercial passenger carriers. For this growing business the builders would need substantial numbers of engineers with professional training. About 1926 several engineering schools expanded their academic programs of aeronautical engineering, mainly thanks to the largesse of the Guggenheim Fund for the Promotion of Aeronautics. Two of these—MIT and New York University (NYU)—had especially strong ties to the industry, and so their situations provide interesting background and contrast to that of Caltech.

"The alliance of industry with science is just beginning."[12] With this heady declaration New York University Chancellor Elmer E. Brown broke the ground for the Daniel Guggenheim School of Aeronautics in October 1925. He was but one of many technocrats who envisioned the strengthening of that alliance in America. To foster it mining industrialist Daniel Guggenheim would endow aeronautical departments not only of NYU but of six additional schools in the next four years: those of the California Institute of Techology, Stanford University, University of Michigan, MIT, University of Washington, and Georgia School (later Institute) of Technology. All save Caltech were geared to produce working engineers more than scientists, and each was chosen for its expertise and its representation of a particular region of the country. They constituted every major program of aeronautical engineering in the country in 1929.

Because MIT, and NYU are typical of the institutions and their relationships with industry that Guggenheim meant to support, they deserve attention. MIT had offered courses since 1913, when airplane designer A. A. Merrill, later of Caltech, delivered engineering lectures on powered flight. In that year the navy detailed specialist Jerome C. Hunsaker to establish at MIT a full-fledged program of teaching, especially in aerodynamics; a year later Hunsaker had a four-foot wind tunnel installed in the laboratory there. When Hunsaker entered active naval service in 1916, he was succeeded at MIT by Alexander Klemin, who in turn passed the chair to E. P. Warner upon entering active service a year later. Warner directed the aeronautics department of MIT into 1926, when he was named assistant secretary of the navy for aeronautics, one of three new cabinet air-posts created at the recommendation of the Morrow board.[13]

Throughout its early development the MIT department maintained a balance between teaching and research, underlining in public the need for basic research in tandem with producing engineers. Protectiveness of academic freedom showed through MIT President Samuel Stratton's emphasis, in a report to the Guggenheim Fund, on pursuing *general* problems rather than specific ones such as Stanford had chosen in its propeller research.[14] The Guggenheim Fund had granted MIT more than $200,000 for construction of a new aeronautics laboratory with no qualms about the balance that was proposed. Still, when a litany of accomplishments appeared in an official history of the fund in 1942, it was MIT's impact on the aviation industry that the author cited first after listing the number of degree recipients: "Among its graduates are chief engineers or engineering directors of the Curtiss–Wright, Glenn L. Martin, Pratt & Whitney, Vought, Hamilton Standard, Lockheed, Stearman and Douglas

Companies, as well as engineering officers of the Naval Aircraft Factory and of the Air Corps Materiel Division at Wright Field."[15] Then "contributions to aeronautical science, meteorology and automotive engineering" were cited in a general way, then inventions, then textbooks by the staff. No such listing of engineers ever appeared for Caltech, because in truth, few of its graduates ever entered industry. In the same history, basic research in aerodynamics and structures at Caltech was described in detail, as was airplane wind-tunnel testing.[16] Behind these contrasting references lay two important facts: MIT had committed itself to training engineers and carrying out basically engineering research, whereas Caltech, under the direction of Theodore von Kármán, viewed itself instead as a center of research in "applied science"—a step removed from building airplanes.

In sharp contrast to Caltech, NYU saw its role as even more the handmaiden of industry than did MIT. Recipient of the first Daniel Guggenheim grant in 1925, even before the founding of the Guggenheim Fund, NYU had already begun in aeronautics in 1922. A practical course in aeronautical engineering, organized by Collins Bliss and taught by Alexander Klemin that year, brought favorable publicity to NYU's financially ailing engineering school. "Reports of the Dean of the School of Applied Science" between 1911 and 1917 had shown growing complaints of inadequate facilities and underpaid staff. But a new engineering building appeared in 1920, and the school's name was changed to College of Engineering, clearly to reflect a more practical and industry-supported curriculum. In 1923 Dean Charles H. Snow wrote confidently, "Inasmuch as the principal manufacture of airplanes and airplane engines centers within a comparatively short distance of New York City and so of this campus, cooperation between the University and the Industry is entirely practicable."[17] Robert Millikan would argue identically of The California Institute of Technology and the industry in Southern California. Both claims could be true if one distinguished NYU's claims for the present from Caltech's for the future, though in fact Caltech's relationship to industry was to be very different from NYU's.

Snow wrote further of his school's aeronautics in 1925, "This course has received and is yet receiving more attention from the popular press than any other now offered by the College of Engineering," concluding, "More money is needed."[18] The money would go to establish an endowed professorship and research laboratories in aeronautics. Even before plans for a fund drive materialized, however, Daniel Guggenheim intervened to support the laboratory, and the endowed chair became instead the Guggenheim School of Aeronautics at the New York University.[19]

No clearer statement exists of the goals at NYU that Guggenheim was to support than that in the 1924 bulletin of the College of Engineering: "The spirit of the course at New York University is to train aeronautical engineers rather than aerodynamicists, men who can take part in the practical work of designing and constructing airplanes and dirigibles and their engines on a scientific basis. The purpose is to train men who can further the commercial utilization of aircraft."[20] Of fifteen courses offered in the academic year 1926–27, only two were called "aerodynamics," and even they had far more to do with principles and data of wind-tunnel testing than with mathematical theory. The gift of a four-foot wind tunnel from the Curtiss Company in 1924 had allowed the experimental course to be offered.

NYU Professor Klemin publicly associated the school with plane-building: "one aspect in the training of aeronautical engineering which must not be neglected . . . is contact with reality, and in this instance, flight itself.[21] An advisory board comprising Orville Wright, Sherman Fairchild, Grover Loening, Earl D. Osborn, and Juan Trippe, among others—all men who built and flew airplanes—would help to guarantee it. One final fact, cited by Klemin two decades later in his "Brief History" of aeronautics at NYU, shows that the lessons took: 350 graduates were employed in industry while eleven taught, of whom but five had become professors.[22]

It was against this academic background that Robert Millikan drew his plan for aeronautics at Caltech. The contrast was stark indeed, yet Millikan's plan was like the others in one respect, for he sought a subvention from the Guggenheim Fund to build his center. As the action of a philanthropic foundation, the grant of the award must itself be seen in the broader context of endowment support of science and technology in the twentieth century. Out of that context emerges an interesting connection between general foundation ideology and the Guggenheim awards.

In 1936 sociologist Eduard Lindeman published an impressive, comprehensive analysis of the origin, ideology, and influence of American philanthropic foundations. Of the 100 endowments that he examined (80 private foundations and 20 community trusts, roughly one-third of all endowments of the time), only six had existed prior to 1900; 82 had been founded after 1910. The foundation, with its influence on American culture, is a twentieth-century phenomenon.[23]

Lindeman explained the rise of foundations plausibly as the channeling of industrialists' surplus wealth, because of their new social consciousness (or "guilt feelings"), into activities promoting American education, health, and welfare that would foster American culture. "Education," particularly, had become a symbol of culture and "it became the fashion to call a great many activities by the name of education," according to Lindeman. In the decade of the 1920s, 61 per cent of foundation activities were concentrated in the broad category of support for higher education, in the form of grants to colleges, universities, and institutes of higher learning. Another breakdown for the same period had 43 percent of *funds* going to education, 33 percent to "health (including university medical programs)," and 14 percent to "social welfare." Not only was foundation influence on culture a new phenomenon, but it was highly concentrated in teaching and research at universities.[24]

With the outbreak of world war in 1914, a new alliance between industrialists and scientists in the cause of national defense had brought philanthropic as well as government support for teaching and research in the sciences. Military support effectively ended in 1919, but endowments like the Carnegie Institution and the Rockefeller Foundation continued to apply their resources to physical and medical sciences. Those endowments intending to support science or engineering appointed directors who were industrialists or scientists. The Engineering Foundation (created in 1914), for example, supported engineering research and, after the war, National Research Council activities. Its board included Bell Laboratories head and Throop Institute graduate Frank Jewett, who would also help to bring Guggenheim money to Caltech for aeronautics. The Engineering Foundation was headed by W. F. M. Goss, a former dean of engineering at Purdue. This was the exception, for the industrialists' voices on such boards were stronger and more numerous than were the academics'.[25]

Their philosophy of granting subventions for scientific and engineering research is particularly relevent to Caltech's growth in aeronautics. As historian David Noble argues, the industrialists stressed their self-interest in supporting universities. American Telephone and Telegraph's (AT&T) John J. Carty for one (also a supporter of Robert Millikan's aeronautical plans in 1926) asserted in 1924 that research money "should come from the industries themselves, which owe such a heavy debt to science . . . [and] the time has come when our technical schools must supply in largely increasing numbers men thoroughly grounded in the scientific method of investigation for the work of industrial research." Jewett too "urged that the balance (of industry-supporting research) be shifted back to the universities."[26] The apparent irony of the Bell Labs' head arguing a shift away from his own center back to the universities diminishes when one sets his views in the larger scheme of a grand alliance between industry and the universities.

From this environment emerged Daniel Guggenheim's Fund for the Promotion of Aeronautics. The aircraft industry and other advocates of aviation generally perceived in 1926 that its scientific base was especially weak. Yet airplanes were particularly susceptible to "scientific" analysis, as the European aerodynamicists had shown. Given a predilection for aeronautics such as Daniel Guggenheim's son Harry already possessed, one could see a "clear need" (in the sense of complete agreement with foundation ideology) for such a fund as Guggenheim senior would establish. Created in 1926 shortly after Guggenheim's first award to NYU, with total assets of $2.5 million, the fund was small by foundation standards of the time. But since very little support for aeronautical study was available before it, the fund's influence was seminal. It was the major philanthropy in aviation until 1930, when it was dissolved.

The industrial technocrats who controlled foundation support in this country—Carty and Jewett, Goss and Root, for example—included several who would support Caltech's own alliance of science and industry through aeronautics and who would have influence on the Guggenheim board. Millikan, well connected personally to these, had every reason to expect support for education and research in aeronautics from the Guggenheim fund.

2

FINDING "A SCIENTIST OF ABILITY BORDERING ON GENIUS"

Judging by correspondence and the speed of events, Robert Millikan had in mind an aerodynamics laboratory for Caltech well before he met with Harry Guggenheim in 1926 to solicit its endowment. He credited John J. Carty, chairman of the board of Bell Laboratories, advisor to Guggenheim, and member of the advisory committee of the California Institute of Technology, as "the real initiator of the whole movement" to create such a laboratory.[1] But Millikan's advisor, the physicist Paul Epstein, wrote Kármán that it had been Millikan's plan for years to build the facility.[2] Around 1920 aviation had begun to grow in Southern California, and the technocrat Millikan noticed it, as he noticed the growth of the wireless telegraph, oil, chemical, and other technological industries in the country at large. This industrial expansion rested on engineering achievements characteristic of the era: the radio, petroleum "cracking," synthetic organic processes,[3] and, if less profitable yet no less characteristic, the invention of the powered airplane. These achievements all lent themselves to "scientific" analysis. Aeronautics especially, beset in the past by much crankery and bad luck, deserved such analysis according to many American leaders in the field. Millikan shared their belief, and he knew, furthermore, that the problems of the science of flight had already been attacked. He was determined to add certain applied scientific research in his paradise of cooperative scientific investigation. Aeronautics, specifically the scientific discipline of aerodynamics, was an obvious field to cultivate.

Carty had impressed upon Millikan that the time and Guggenheim's disposition were right for setting in motion the scheme that Millikan had been developing. A few days before the new year, Millikan collected letters of endorsement of his plans for aeronautical research from Harry Chandler, publisher of the *Los Angeles Times* and major financial backer of Caltech, Donald Douglas, the aircraft builder, and the board of directors of the Western Air Express Company.[4] He submitted only the Douglas letter, which was most to the point of his plans for research, with a proposal to Guggenheim. Just six months later, in July 1926, agreement had been reached for a Guggenheim grant.[5] A laboratory was already being designed, and Millikan was asking Theodore von Kármán to cross the Atlantic to review the proposed wind tunnel.[6]

Millikan had traveled to New York to present his proposal to Harry Guggenheim and his senior assistant Hutchinson I. Cone in the first week of January 1926. His ideas were not entirely well received. It seemed, as Millikan related the conversation in a letter to his friend Frank Jewett some weeks later, that Guggenheim had balked at Millikan's repeated references to "research," a word that Millikan acknowledged might be "over-abundant in my vocabulary." Guggenheim then had asked Millikan to change his proposal from a request for a permanent gift of $0.5 million, whose interest would pay salaries of new aerodynamics research faculty, to one for a smaller capital grant for "certain installations," which later would constitute the Guggenheim Graduate School of Aeronautics, and for "certain assistance for a period of say five years in the way of helpers."[7] Harry Guggenheim and his father Daniel wanted Daniel's money turned to small grants and to bricks and mortar for a laboratory; not a single, large perpetual endowment. The issue of the centrality of research remained open, ultimately to be resolved to Guggenheim's satisfaction with the addition of teaching to the proposed functions of the new school.

Millikan's meeting with Guggenheim took place before the creation of the Daniel Guggenheim Fund for the Promotion of Aeronautics, whose founding on January 16, 1926, hailed Guggenheim's large-scale commitment to support American aviation and its development. In early February, president of the new fund Harry Guggenheim and his vice president Hutchinson Cone would begin a three-month tour of Europe seeking ideas for the direction of the fund in the aeronautical centers of research in England, France, Germany, Holland, and Spain.[8]

Millikan had been dismayed by Harry Guggenheim's response to his initial proposal, for he felt that the lack of funds for hiring researchers would prevent his establishing a "real center" of aeronautics of the quality of Caltech's departments of physics, chemistry, and mathematics. This high quality would be the hallmark of aeronautical research at Caltech. Millikan wrote to Jewett to ask if Jewett could not help "get these ideas into Mr. Guggenheim's mind and into that of his father." He suggested Jewett talk with his old supporter Carty in New York about the matter of permanent grants for salaries, for Carty was close to both Guggenheims.[9] The same day Millikan wrote to friend John C. Merriam, president of another foundation for scientific research, the Carnegie Institution of Washington, to ask Merriam "to advise Mr. Harry Guggenheim or his father to this [same] effect."[10] Their efforts may have helped, for when negotiations were concluded, the Guggenheim Fund had agreed to support some research salaries, if only for ten years.

In Millikan's original "Proposal for a Research Center in Aeronautics at the California Institute of Technology," he noted that there were three outstanding centers of aeronautical progress in Europe. The French Aerodynamical Laboratory in Paris continued "remarkable work started by Eiffel," which, begun around 1900 with measurements of the air resistance of a flat plate dropped from the tower of his name, now embraced sophisticated techniques of aerodynamic testing in Eiffel's wind tunnel.[11] The

National Physical Laboratory near London was noteworthy for supporting seven "mathematical physicists" concerned with the theories of lift and drag. Among them Millikan listed H. Glauert, R. Jones, and one G. I. Taylor, whose name would reappear in his letters. But it was Göttingen that Millikan mentioned first: "One [outstanding center] has been in the famous little university town of Göttingen, Germany, where Prandtl, a well-known mathematical physicist, who before he took up aerodynamics had made a world-name for himself in the field of hydrodynamics, has since 1908 developed the theory of the forces on the airplane wings which has thrown a flood of light on the underlying principles of aeronautics."[12]

Millikan was reflecting general belief among those who knew aerodynamics that there were few centers of excellence in mathematical aerodynamics, and that none were in the United States. But he had special reason to observe this in his proposal, because he now claimed that American weakness in aeronautics was the result of this lack of aerodynamic science in a center of both theoretical and experimental research. "Indeed, this country has been very weak in centers of mathematical physics, and weakness in aeronautics has been a natural consequence." The key, he claimed, lay in strengthening mathematical physics. He would build a center at Caltech, which would incidentally attract the resources of the growing aviation industry in support of his new institute. Mathematics and physics already flourished at Caltech, thus strengthening his case for a center of mathematical aerodynamics. Here, he said, the chances of advancing aviation were exceptional.

Robert Millikan, himself an experimental physicist who had won the Nobel Prize in 1923, counted several mathematicians and physicists as outstanding members of the California Institute of Technology, among them Harry Bateman, Paul S. Epstein, Richard Tolman, and E. T. Bell. In his first proposal to Guggenheim he chose to emphasize this present strength and others of Caltech's, such as its proximity to Douglas's aircraft plant and its "climatic and topographical advantages." He did not focus on how a grant for salaries would alter the course of aeronautics in America. To be sure he had appended a four-page "Program of Research" to the five-page proposal, which included a list of fourteen investigations that were being pursued at the moment. But these by no means added up to a revolutionary, hardly even an accelerated, program of research or of education. After his conversation with Harry Guggenheim, the emphasis changed.

At the end of January, Millikan sent a "preliminary letter" to Guggenheim continuing his argument for a grant to Caltech.[13] Millikan underlined the two "quite unique" advantages of Caltech, which "constitute far and away the most essential elements in a program of aeronautical development in the United States"—the fact that the Los Angeles area was already a center of production of aircraft, and "the existence here already of conditions in mathematics and physics which make the Institute altogether unique for a man seeking to do Ph.D. work in the field of hydrodynamics and aeronautics." He knew of Guggenheim and Cone's imminent travels to Europe and offered his aid in gaining them access to information, especially in Germany. He might be able to help "because of the fact that Kármán, who is probably the most competent man in Europe in the field of aeronautics, is a friend of our Dr. Epstein at the Institute and also well known to Dr. Bateman." (The mathematical physicist Paul Epstein had attended professional congresses with Kármán, and Harry Bateman had helped develop the field of viscous fluid flow, to which Kármán too had contributed.) Caltech's new laboratory would need a research leader, and Millikan was suggesting that Guggenheim, on his trip to look over the men and their institutions, should not be deterred from examining the Germans as well as the former allies. Millikan offered also to send a representative of Caltech with Guggenheim or to have them meet in Germany. "One of our ablest young mathematical physicists is now in Göttingen and might be of much use in connection with the visit to Prantl's [sic] laboratory, while the association of our group with Kármán at Aachen would also open a door which might otherwise remain closed." This would ensure that Guggenheim examined Göttingen and Aachen, the first of which Millikan knew to resemble the center he wanted to create. In addition, Millikan was no doubt attempting to consolidate his position with Guggenheim by offering a favor and by seeking a further chance to convince Guggenheim and Cone of the merits of his plan.

Harry Guggenheim had already left when Millikan's letter arrived in New York. Hutchinson Cone responded from New York just before he too departed. He declined with thanks Millikan's offer to send along a Caltech representative, commenting pointedly that such an errand would not further the interests of the California Institute. But he would appreciate entrance into laboratories, especially those in Germany and particularly Kármán's.[14] Evidently foreign scholars and entrepreneurs had not easily gained admission to German aeronautical laboratories in the past, possibly because of concern over international patent rights, more likely because of lingering nationalistic animosities and German sensitivity to allied "inspections," and in order to conceal research that might be construed by the allies as bearing military significance.[15] Although some of these reasons did not apply to Kármán, he had done considerable work for German industry and the military, as had Prandtl.[16] For whatever reason, the Guggenheim tour failed to visit Kármán. Prandtl, however, was interviewed at length, and the scope and quality of his activities made a deep impression upon Guggenheim.

It is difficult to know if Harry Guggenheim believed when he left America in the importance for Caltech of a strong director of research, one who would set the tone of the laboratory and lead assistants and graduate students in the directions he judged significant. But if Guggenheim had no such conviction then, his mind was made up by what he saw in Europe and especially in Germany. In a report of his travels to the Guggenheim Fund, he emphasized the importance of strong, central leadership. Noting that European experimental facilities were not as good as those in the United States, he observed that out of them, nevertheless, came more and better scientific research and workers. Efficient organization and brilliant direction were the keys to their greater productivity:

Europe has few splendid laboratories and equipment for scientific research in aeronautics such as exist in the United States. And nowhere in Europe are necessary funds available to carry on important and vital research. However, the countries of Europe are far richer than the United States in highly trained scientists who are devoting their lives to aeronautical problems. Professor Prandtle at Gottingen [sic], Germany, is recognized and in fact revered throughout Europe as the foremost scientist in Aerodynamics. He carries on a large part of the important fundamental aeronautical research work in Germany with the assistance of post graduate students who come to study under his guidance. In this manner he has developed and continues to develop highly trained research workers. *The United States would seem to have the need of such a center. This is difficult of attainment because its success would be dependent upon the scientist about whom such a center were created.* [italics added][17]

A likely place for such a center was Caltech, for it had above all else the will to become a nationally recognized center of research in aeronautics. It had also the energies and insight of Millikan, who, Guggenheim knew, had already succeeded in acquiring outstanding leaders in research. Among thirty-five suggestions in his report for proper use of his father's endowment, Guggenheim listed as number 5, "Endeavor to locate a scientist of ability bordering on genius and place at his disposal facilities and opportunities to enlarge his usefulness," and as number 6, "Establish a

laboratory or center for the investigation of revolutionary ideas."
Caltech would be the closest thing to it that Guggenheim would
found.

Millikan arranged to meet Guggenheim again on his return
from Europe.[18] Discussing the evolution of their ideas about the
center at Caltech, the two agreed that the original proposal should
be submitted to the trustees of the fund, but that it should be
amended by an accompanying letter. The letter was to contain a
"condensed reformulation of the conditions at the Institute" that
would foster such a center, to reinforce the case for Caltech, and
specific, new proposals for building a laboratory. In the letter
submitted in May, Millikan reemphasized that Caltech was
"unique in the number and calibre of the mathematical physicists
associated with it," among them the four already mentioned, who
constituted an extraordinary combination of talent in mechanics
and aerodynamics.[19] He also noted with enthusiasm the "very
large amount of work going on here in the allied fields of physics,
chemistry, mathematics, astrophysics, radio, and all branches of
engineering," thus underlining the importance of creating an atmosphere for cooperative research in related fields. It was a theme
that pervaded his vision of Caltech and that had gained him financial and institutional support for other departments—of physics,
chemistry, mathematics—as well.[20] He proposed that Guggenheim grant funds to develop four aspects of a new center.
The Daniel Guggenheim Graduate School of Aeronautics, a research and teaching center, would require a new building with a
wind tunnel of test section ten feet in diameter (a large dimension
for the time); it would need endowment for support of a scientist
of Bateman's quality; and it should have an annual appropriation
of $10,000 for ancillary facilities and expenses of research. But
installations now took priority.

In two weeks the trustees of the Guggenheim Fund approved
Millikan's proposal. The majority felt that funds for salaries
should be granted for only a limited number of years, a trial pe-

riod, because "aeronautics is a new art."[21] Harry Guggenheim, whose suggestions in these meetings not surprisingly were almost always heeded, had sponsored both the proposal and the restriction. The latter was in keeping with the Guggenheim view that a grant should be seed money and not ongoing support. The salary for a scientist was limited to $10,000 per year for ten years, but the fund awarded $180,000 for building. The research allotment, at $5,000 per year, was half that requested and granted for five years only.

Stanford, which until May had not figured prominently in the negotiations, received at the same meeting $195,000 to build its program of education in aeronutics. Guggenheim had supported Stanford's and Caltech's applications with the following statement: "At the present time there is no educational center in aeronautics on the West Coast of America. Here would seem to be a real need for such a center now and by the time it could be established and results obtained, no great imagination is necessary to predict that the need will be vital." He recommended the establishment of this center in the West at "the most appropriate existing institution." Education, Guggenheim noted, was "like an investment in gilt edge securities."[22]

Millikan lost something, both to Guggenheim's ideology and to Stanford's claims, but he achieved most of what he sought. He had, in fact, worried that he would lose more, and he had taken steps to enlist the support of friends who might influence the decision of the trustees of the Guggenheim Fund. Three days after he had forwarded his proposal of May 14, Millikan wrote the letter to Carty mentioned at the beginning of this chapter. There he assured Carty that aeronautics at Caltech would start along proper lines of high-quality research. The work would be "altogether comparable with that which we are doing in physics, chemistry, and mathematics." Millikan felt that Cone and Guggenheim were "altogether favorable . . . unless some members of the board throw cold water on the enterprise." He mentioned William F. Durand, senior board member from Stanford who believed that his institution as well had the basis for a claim. Evidently Durand had expressed his belief to Millikan, who feared that a competition between the Institute and Stanford would lead to no action at all. The case for Caltech was strong, he asserted, but the problem lay in a "weak spot in the psychology of the situation." A certain deference had to be paid Durand; still he could be reasoned with. "Dr. Durand is himself a broad-gaged man and a very intimate friend of mine. Further, I think he would himself see the danger of actually interfering with a very important aeronautical development if too strenuous a debate regarding the merits of different localities is entered into. If such a debate does arise I want very much that somebody who is conversant with the situation here have a hand in the presentation of this case"—by which he clearly meant an advocate of Caltech. Millikan asked if he could enlist Carty to "get a line from Harry Guggenheim himself, or Admiral Cone, as to who is likely to be present at the board meeting, and in case Dr. Durand or Professor Michelson are to be there if you could perhaps have a word with them beforehand which might have an inhibiting influence upon any actions which might lead to unfortunate results for the actual development of aeronautics in the country." If he was unable to present his case in front of each member of the board, he would delegate Carty, whom he knew to share his motives, to present Caltech's case. With all proper gentility, of course: "I am sure that a word from you to Guggenheim and perhaps to some of the other members of the board would be of immense importance at this juncture."[23]

Millikan had mentioned his elder colleague A. A. Michelson, the Nobel Prize-winning physicist at the University of Chicago, with good reason, for he knew him to be sympathetic to an award to Caltech. Evidently Millikan had written to Michelson, who was also a member of the board of the Guggenheim Fund, at about the same time as to Carty. For on June 3, one day after the board had passed its favorable resolution (unbeknownst to the inattendant Michelson), Michelson forwarded to Millikan the draft of a letter of endorsement that he was prepared to send to Guggenheim upon Millikan's approval. Michelson would conclude, "If anything which I have said would make it likely that the Board would take the matter [of Caltech] into favorable consideration I should feel that the selection of my name on your board will have been justified."[24]

We can assume that such a letter was never sent, because Millikan received notice of the resolution June 7.

Perhaps through the good offices of Carty acting as Millikan's representative, out of the generous resources of the Guggenheim Fund, both Caltech and Stanford received substantial grants. Since Stanford already possessed an aerodynamics laboratory, its grant amounted to less. The money was designated for improving the laboratory and its building and for creating a six-year undergraduate course in aerodynamics and aeronautical engineering. This course would qualify students "for effective service in the aircraft industry and in air transportation activities, where there is promise of early and large development."[25] Caltech, the other recipient of Guggenheim's first West Coast gift, would become the center of academic aerodynamics, especially in its mathematical aspects, led by the "scientist of ability bordering on genius" Theodore von Kármán,[26] though Kármán did not yet know it.

Millikan announced in Pasadena in August that the funds provided would make possible six different activities in aeronautics at Caltech.[27] The first was the "extension of its theoretical courses in aerodynamics and hydrodynamics, with the underlying mathematics and mechanics, taught by such men as Professors Harry Bateman, Edward T. Bell, and Paul S. Epstein." Following this were the initiation of practical courses taught jointly by Caltech's experimentalists and by the engineering staff of the Douglas Airplane Company; a comprehensive research program in airplane and motor design; the perfection of a new tailless airplane, which constituted "a radical departure from standard aeronautical design"; several research fellowships in aeronautics; and research on full-scale experimental aircraft as well as on wind-tunnel models. From events to come and from the expressed intentions of Millikan and Guggenheim, it soon became clear that the first item—the teaching of theoretical aerodynamics (and its associated pursuit in research)—would be the primary locus about which the new graduate school should develop. It was the field that Prandtl and his students dominated in Germany.

Although Harry Guggenheim had underlined the significance of Prandtl's work in Göttingen, the flourishing of aerodynamics in that Prussion university began with the organizing efforts of another. Felix Klein, the mathematician, had bequeathed to both Prandtl and Kármán his belief in the utility of allying mathematics and physics with engineering, and he had put that belief into practice by establishing between 1896 and 1904 three new chairs of applied science at Göttingen, one of which became Prandtl's.[28] In that Caltech's new School of Aeronautics would owe much of its excellence to the products of the aerodynamics institute at Göttingen, which is suggested by Guggenheim's report to the members and confirmed by Kármán's coming to Caltech, it is useful to look at the reasons for the strength of Göttingen, and especially at Klein, Germany's early and foremost advocate of applied science in universities.

3

APPLIED SCIENCE AT GÖTTINGEN

Only about the turn of the century did the small group of American physicists emerge as a professional community, Daniel Kevles has argued in his book *The Physicists*. Together they would seek to create a "sustained and outstanding indigenous physics" through "the setting and enforcement of standards of excellence."[1] This meant the systematic exclusion from professional journals and activities of amateurs, hacks, and inventors—among whom, in the last category, were some of the most respected engineers. Although there was by no means unified opposition to the pursuit of applications in this new profession, as for example at the U. S. National Bureau of Standards where establishing benchmarks for new technology remained the reason for pure research, a heirarchy of research developed in which the most worthy pursuits were labeled "pure." Men like inventor-entrepreneur Thomas Edison were often excluded from the physics community, and they in turn shunned the more abstruse forms of science or engineering that involved mathematics and theory. Even the terms *pure, applied, basic, theoretical*, and *practical* carried connotations of value—as they still do—that meant exclusion from or inclusion in the community of physics. An idealized physicist, a pillar of the community, was on the whole someone who pursued "pure" or "basic" experimental research grounded in a theoretical foundation (that had usually been created in Europe), who qualified for the title by contributing to the literature of the profession, and who protected the integrity of his profession by seeing to it that standards were upheld, in part by excluding those who merely pursued applications or practical ends.

Yet earlier, in the mid-1890s, Felix Klein in Germany had established alliances of science, government, and industry that led to the incorporation of engineering studies into scientific faculties at German universities, Göttingen in particular, and the consequent creation of an academic class of researchers pursuing *applied science.* The term (*applied mechanics* more particularly), like the activity, was a hybrid. It carried the senses of both knowledge for use and knowledge for its own sake, and its invention suggested that the two were not mutually inconsistent.

The idea of allying scientific and technological research in Germany was an old one. One historian of science has written that "German physicists organized themselves into a discipline-community at the time [c. 1870] that German society was with unprecedented singlemindedness bent on economic, cultural, and political aggrandizement through the intellectual and material domination of nature."[2] German physical research, even "pure" research, had benefited from society's pursuit of this domination, for it thereby had received generous grants. But, the same author notes, "the locus of technical education had shifted from the universities to special schools," the trade schools and the polytechnic institutes, which grew to importance in the German educational system in the latter half of the nineteenth century. In 1887, a partnership of Hermann von Helmholtz, backed by the Berlin Physical Society, and wealthy industrialist Werner von Siemens brought to completion the Imperial Physical Technical Institute, a prestigious research laboratory roughly analogous to the later U. S. National Bureau of Standards and created to investigate both physics and technology. But this laboratory remained outside the German educational tradition of the universities, which for the most part by 1895 prized their freedom from pursuing "applications."[3]

About the same time, however, Felix Klein (figure 1) took steps to further research linking the disciplines of science, mathematics, and engineering at the University of Göttingen. Klein, who was born in 1849, had arrived in Göttingen in 1886 with a supreme reputation in mathematics, after holding professorships at the University of Leipzig and earlier the Polytechnic Institute of Munich and the University of Erlangen, where he had received his first professorial chair at the age of twenty-three. It is sometimes remarked that Klein's idea of bringing mathematicians to look more favorably upon applying their discipline derived from his observing university programs in (civil and mechanical) engineering in the United States during travels there in the 1890s.[4] Such programs were unknown in German universities. But it came as much from his own pragmatic mathematical tastes, now turned to teaching and organization of research.[5]

Klein's conception of applied research had two salient aspects,[6] which were reflected in the functions of the two chairholders of the Division of Technical Physics, established by Klein at Göttingen around the turn of the century. The occupant of the post that evolved to the professorship of applied mechanics (first Eugen Meyer, then Hans Lorenz, then Ludwig Prandtl from 1904 to 1947) used his abilities in mathematical analysis to solve problems of general engineering interest. In Prandtl's case these problems were such as the stability of elastic structures and the flow of fluid past an obstacle. He was thus much a theorist applying his science to problems of practical significance, given also, however, to their empirical investigation. Today such activity might be called "applied science," or more restrictively (in the terms of historian Edwin Layton) "engineering science" to distinguish it from "engineering design."[7] The distinction between engineering science and design or development was important even before 1900, for it came to be the basis upon which Klein argued that German universities should pursue applied science in general. Klein had largely to *exclude* design from his proposed program of research (even if it crept in later), ceding this aspect to the *Technische Hochschulen*, or polytechnic institutes.

Carl Runge, the occupant of the professorship of applied mathematics, conducted a more ideosyncratic enterprise, applying his mathematical skills, particularly numerical calculation, to physical problems of the purer kinds. Runge, who came with Prandtl from the Polytechnic Institute of Hanover to Göttingen in 1904, had applied his mathematics to discern series formulas from the prodigious amount of spectroscopic data that he had amassed with Heinrich Kayser and Friedrich Paschen before 1900. For him applied mathematics meant also the pursuit of physical experiment. In fact Runge came to Göttingen expecting to concentrate his energies in experimental work on the splitting of spectral lines in an external magnetic field, the phenomenon called the Zeeman effect. This conception of his role, though largely Runge's own, received the understanding and support of Klein. (Observing that he had believed for years the prevailing methods of asking mathematical questions to be "artificial," Klein wrote Runge in 1896: "You know how much I value the form of posing questions with which you are occupied."[8]) Runge retained Klein's support in ensuing years by emphasizing the teaching of applied mathematics in his texts and courses and by endorsing Klein's calls for reform of mathematical curricula—both purposes that were dear to the older mathematician.[9]

So the centers of gravity of the two chairs rested at different loci of "application," but the posts covered considerable areas. In the direction of practice, applied mathematics approached theoretical and experimental physics, whereas applied mechanics approached engineering. At the "purest" extreme, the former reduced to methods of computation, the latter to theoretical physics. There was thus an intersection of the two chairholders' fields of interest, located roughly at the foundations of mechanics. This and Klein's support of the full spectrum of activity suggest, moreover, that Klein's advocacy of applied science did not depend on such distinctions as between pure and applied science or between mathematical and experimental methods. Rather he had seen a general trend in mathematics and physics toward the abstract.[10] Against this he revolted, sparked by his intellectual conservatism, which coincided with (if not arose from) his broader social attitudes. Preferring the concrete, the tangible, the pragmatic, he filled each post in applied science with men who shared his view of the proper direction for science.

Yet it was not only his taste for intellectual advancement but also a nationalistic zeal that drove Klein's pursuit of applications. For in his plan to make Göttingen the world center of "technical physics" he sought the larger goal of putting German science, with Göttingen at its center, in the lead of all scientific progress. A telling sign was the seal of Klein's "Göttingen Association for the Advancement of Applied Physics and Mathematics" (figure 2).[11] Encircled by the words "Göttinger Vereinigung" and "1898," the year of its founding, the figures of Mercury and Minerva stood together clasping the trunk of a tree bearing golden fruit. Behind these figures, standing on the summit of the near-sacred Hainberg mountain, rose the spires of Göttingen in the distance. What more proper symbol for an alliance of German commerce and science, steeped in their Holy Roman heritage, nurturing the fruit of technological progress.

Göttingen's history proudly recalled the efforts of Carl Friedrich Gauss and Wilhelm Weber to collect and use precision instruments. Gauss especially, who spent his entire adult life in physical and mathematical research at the Georg-August University, set an example for the applied mathematicians whom Klein would encourage almost a century later. At home in the statistics of data reduction, which he largely invented, as well as in the more abstract fields of number theory, analysis, algebra, and geometry, which he revitalized, Gauss applied many of his mathematical discoveries to empirical and theoretical investigations of astronomy, surveying, geomagnetism, and electromagnetics. In this career tour de force, Gauss managed to equip the Göttingen astronomical observatory and to sustain a fifty-year interest in observing. He developed the heliotrope, an instrument to provide an intense source for surveying by reflecting a beam of sunlight. He invented a working telegraph with young Weber in 1833. He developed various magnetometers, and he built a magnetic observatory at Göttingen.[12] His successful amalgamation of experimentalism, invention, and rigorous mathematical analysis in his own career was to be in great measure Klein's ideal of applied science. No wonder that Klein labored for decades to edit and publish his papers, for Gauss's eclecticism in science, especially his fruitful pursuit of applications, had to be appreciated by and encouraged in future generations of Göttingen students.

Weber too connected mathematical method with experiment, though in his career the latter bore the emphasis. He, like Gauss earlier, directed the astronomical observatory; but most of his research lay in the areas of geomagnetism and electrodynamics. He too invented several instruments, including one for measuring electrical currents. He showed through a simple theoretical argument how such an instrument could be used to measure electrical resistance as well. Weber's theoretical musings were predominantly grounded in his own experimental programs at the university.[13]

Thus at Göttingen a venerable tradition of applied mathematics and physics had existed.[14] Perhaps before most other universities, a scheme here for broadening mathematical studies to include physics on the one hand and aspects of engineering on the other might reach fruition, if the arguments and the ground were carefully prepared. For all these reasons Klein sought and found opportunity to further his project of establishing an exchange among mathematics, physics, and practice at his university, while being able to remain largely above reproach, if not immune to resistance, for doing so.[15]

Klein's ideas were not received without controversy. Many

professors opposed the degradation of pure scientific research through taint of material purpose. Klein's habit of arranging for honorary doctorates to be awarded to major, sometimes disreputable, financial donors to the university (a practice common also to American universities even today) gave his argument that an industrial connection would uplift pure research the ring of disingenuousness.[16] Professional engineers resisted what they saw as Klein's expropriation of engineering studies from the polytechnic institutes, for they resented the scholarly contempt for such studies in universities.[17] But German engineering educators withdrew their objections in 1895 with the so-called Peace of Aachen, an agreement among Klein, imperial officials, and engineering leaders representing the polytechnics. It stipulated that the polytechnics would be left to themselves to educate engineers and their own teachers, while universities would pursue "engineering science" (the phrase was theirs) only insofar as necessary to "accommodate the requirements for a versatile education of mathematicians and physicists, above all the future teachers" in the universities.[18] This delimitation of responsibility proved workable, for it defused opposition of the engineers to Klein's grandiose plans while setting forth a higher goal of studying applications in academe that was acceptable to the university professors.

As *Geheimrat*, holder of the title of academic privy councillor bestowed by the Kaiser, Klein was one of a small group of leading German technocrats working on behalf of Göttingen and greater Germany. Energetic, indomitable, and politically astute, he succeeded in organizing seminars and influencing faculty appointments to emphasize technical physics.[19] He brought the greatest talents in these fields to Göttingen. Among them were the aforementioned Meyer, Runge, and Prandtl, electrotechnician Hermann T. Simon and the physical chemist Walther Nernst. Even in the pure sciences, Lewis Pyenson has noted, he exerted an influence on the works of mathematicians David Hilbert and

Gustav Herglotz, of mathematical physicists Max Abraham, Gunnar Nordström, and Max Born, and of geophysicist Emil Wiechert: "Their emphasis on mechanics seems to have been related to the strong Göttingen program in technical physics and to Runge's and Prandtl's research on engineering problems associated with hydrodynamics."[20] But it was Klein who fostered that approach, who hired Runge and Prandtl and then most actively cultivated their membership in the society of Göttingen mathematicians. Richard Courant, the illustrious student of Hilbert's, also observed this convergence of pure research and applications and remembered it years later. As his biographer related, "This general interaction of theory and practice was always to be for Courant the distinguishing characteristic of the Göttingen scientific tradition."[21]

Klein contributed much more along these lines. He advocated not just an alliance between pure and applied science but also a bond between academic mathematical research and industry. He instigated the establishment of the private foundation of industrialists and professors called the Göttingen Association of Applied Physics and Mathematics, by which he fostered the introduction of technological study at Göttingen through grants for chairs and laboratories.[22] Why he would want industrial financial support is self-evident; more marvelous is how he was able to convince German industrialists that they had something to gain by the association. Surely his arguments appealed to those economic, political, and cultural aggrandizements that characterized the period. As with Robert Millikan two decades later, who in other ways invites comparison with Klein, energy and enthusiasm played a major role in his success; and Klein argued with evident conviction that an "industrial connection would be of equal importance for the discovery of new mathematical truths."[23]

He accomplished his task of persuasion by gradually gathering support from those in positions of power whom technical businessmen like steel magnate Krupp and Siemens understood and trusted. First Klein won the confidence of Friedrich Althoff, the ministerial director of the Prussion Ministry of Culture, a man who advanced the cause of technological education in Germany from 1882 to 1907 by supporting unprecedented growth of the polytechnic institutes.[24] Klein had received a tentative offer of a professorship at Yale on his American tour of 1896. It was not the first such offer; he had rejected a similar offer from the Johns Hopkins University twelve years earlier.[25] But now he declined the post, as he noted pointedly in a letter to Althoff, to set in motion his plans for garnering support from industry to bring technical education to German universities. (Klein sought Althoff's support, for the latter controlled both government purse strings and the official sanction needed to advance Klein's proposals.) Althoff, it seems, was impressed by Klein's deference to Germany, and he was won to the enterprise not only by its own attractions for him but also by Klein's evident zeal.[26]

Althoff and Klein tried to bring Klein's former colleague Carl Linde, professor of applied physics at the Munich Polytechnic and developer of the technique and industry of refrigeration,[27] to direct Klein's proposed new "physical-technical institute" in 1896. Linde declined to move from Bavaria but became convinced of the value of the undertaking. In his turn Linde interested Henry Theodor Böttinger, director of the new but economically influential Farbenwerke in Leverkusen of the giant Friedrich Bayer chemical company. Böttinger had encouraged the founding of Walther Nernst's Institute for Physical Chemistry and Electrochemistry at Göttingen in the same year and then came into contact with Althoff. From Linde, however, whom he knew well already, he heard a convinced endorsement of Klein's ideas, from the standpoint of one who had succeeded in both academics and business. Geheimrat Böttinger brought to the growing efforts his prestige in industry, which was as great as Klein's was in academe, and his influence in the Prussian ministries and commissions of the state legislature (*Landtag*).He also gave Klein his first close-up view of the world of business. As Karl-Heinz Manegold, who described the founding of the Göttingen Association in his monograph, has related, "It was through Böttinger that Klein first gained real insight into the world of industrial enterprise, until then so foreign to the university men of learning, which, as the complement of his scholarly field, had always greatly attracted him." A fourth member of the founding group was the Prussian financial administrator, or *Kurator*, of the university (and former councillor to the Ministry of Culture), one Ernst Höpfner, also a friend of Althoff's.

With these powerful converts and donations from Böttinger, Linde, and a locomotive manufacturer named Kraus from Munich, Klein established a technical division of the Physical Institute at the University—the first step of his plan to build an autonomous technical institute within the larger institution. According to Manegold, its opening in 1896 marked the first time that industrial means had been applied to technical higher education in Germany. The founding of the Göttingen Association followed in February 1898, and it was advanced in great measure by efforts of Böttinger, who placed the "scholarly study of great technical problems" at the center of his appeal to fellow industrialists for their financial support. Judging by an account of the first meeting of the founders, we note that they accepted, with some slight reservations, its goals of fostering both teaching and research in such problems. Klein shared with chairman Böttinger in directing a commission of professors, the representatives of the academic members of the association.[28]

Manegold quotes the founders appealing for support of research at Göttingen. Their purpose was to foster a general alignment of "scholarly research and practicing technology." As Paul Forman has pointed out, that sort of alignment materialized largely in the erection of technical institutes like Nernst's and Prandtl's, which were conceived to study problems of a combined scientific and technical nature.[29] The fruits of the Göttingen Association therefore were centers of applied physical research, of scientific-technical knowledge rather than engineering development per se. Engineering teaching, furthermore, was still to be left to the polytechnic institutes, that parallel German academic system.[30] Anton von Rieppel of Nuremberg Machine Co. (MAN) represented the views of the industrial members toward projects of the Göttingen Association when he suggested that it would be well if through the work of the association the university finally would produce "understanding of and interest in technical and industrial questions."[31] Evidently professors had failed in this regard, but now industry would have a chance to excite Göttingen to its needs.

4

AERODYNAMICS AS AN APPLIED SCIENCE

The cultivation of aerodynamics at Göttingen was essentially Klein's doing, yet with the full support of his technocratic alliance. Having established a Professorship of Machinery about 1898,[1] occupied first by Meyer and then by Lorenz, Klein was presented with a special opportunity for advancing his program of "technical physics" by resignations of Lorenz and another professor in 1904. He filled both posts with men interested in aeronautical issues as in other high technical developments of the time. Lorenz was to leave the first chair for a post in the polytechnic in Danzig, and a successor had to be chosen. A. Stodola of Zurich, the premier European professor of applied mechanics, headed the philosophical faculty's tentative list of candidates.[2] But Klein, Böttinger, and Naumann of the Finance Ministry, the man who controlled state salaries for Göttingen, soon realized that his price would be "out of the question." The appointment would be a junior ("extraordinary") professorship anyway, and Stodola would surely not accept such a low post.[3] They fell back, enthusiastically nevertheless, on Klein's second choice, the financially acceptable, highly promising Ludwig Prandtl, age twenty-nine. In fact, Klein had been considering bringing the much-praised Prandtl to Göttingen for four years.[4]

Before negotiations were concluded with Prandtl, the second post opened when a senior mathematician named Schilling announced that he too was heading for Danzig. Klein greeted the news with elation: "The question is if it would be possible to use the advantage gained by Schilling's departure to call *Runge* to Göttingen!" he wrote confidant Julius Elster, with whom he had been sharing his plans for the Göttingen program of technical physics.[5] Continuing, he observed, "the matter depends on Prandtl's call, for Runge is a close friend of Prandtl's." Carl Runge would occupy a new chair of applied mathematics, converted from Schilling's conventional mathematics post. With Prandtl filling the lesser position, the two would constitute a Division of Technical Physics of the old Physical Institutes.[6] By the end of July 1904, despite major efforts by the faculty of Hanover to keep Prandtl,[7] both had accepted appointments.[8]

The Polytechnic Institute of Hanover incurred two serious losses at one time in the move of its professors of applied mathematics and mechanics. Prandtl, whose appointment to Göttingen was confirmed about a month before Runge's, could conclude the financial negotiations only after his successor was chosen. The resistance in the Finance Ministry had as much to do with his financial demands as with his succession. Both were resolved by the end of June.[9] That he was willing to exchange a full professorship (and one with which, he said, he was more than satisfied) for the junior post at Göttingen suggests the great appeal that the situation there held for him. Asked to write his thoughts about the possibility of the move, Prandtl had declared to Klein: "On the one hand I am attracted by [prospects of] my own laboratory and greater free time, but not least by the fine Göttingen scientific relations." On the other hand, he had grown "very fond" of his activities in Hanover, and he would have to exchange a large working group for a smaller one in Göttingen. He diminished the significance of the nominal lowering of position: "I would count it less important that in Göttingen as 'extraordinary professor' I would have no seat on the faculty while here I am an ordinary member of the division." His "weightiest concern" was that he be allowed to pursue his own lines of research, to keep in touch with engineering practice and allow him at some future time to return to a polytechnic institute. He underscored his "feeling of belonging to technology," his "cherished idea" of long standing to be a part of the infusion of scientific thinking into engineering teaching.[10] Given Klein's philosophy of education in the physical sciences, Prandtl's declarations did not hurt his standing with Klein. Indeed, most of what he wrote seems designed to appeal to Klein. Prandtl closed with an accounting of his financial situation at Hanover and his conditions for moving to Göttingen.

He managed finally to emerge from negotiations with his total income improved a bit,[11] but more important to him no doubt was Klein's offering the promise of a research laboratory and probably a future ordinariat, both of which Prandtl had by 1907.

In 1905 the physics laboratories were reorganized and enlarged as part of Klein's broader program for expanding sciences at Göttingen. A modern Physical Institute for the two Divisions of Experimental and Theoretical Physics, the sections devoted to pure research, opened in early December. The Institute for Applied Mathematics and Mechanics, independent successor of the former Division of Technical Physics, retained the old physics quarters. Runge (figure 3) and Prandtl (figure 4) shared for a time in directing this new institute.[12]

Until now Ludwig Prandtl's research had encompassed problems of mechanics of solids (especially the theory of elasticity and "critical phenomena") and of flow of liquids, but not of gases. He would soon widen his area of concern into aerodynamics. In Prandtl's obituary of Klein, he credited his patron with introducing aeronautics to Göttingen: "It was solely his idea that the science of flight be pursued at Göttingen."[13] In 1906 a Society for the Study of Motor Airships was founded with the endorsement of the kaiser. As a foundation for technical research, its purpose was to support aviation, especially the development of the nonrigid Parseval airship, through private donations.[14] Its members would include, among early leaders of German aviation, the rival airship developers Ferdinand von Zeppelin and August von Parseval, as well as German industrialists and bankers. The same Althoff, representative of the Ministry of Culture, supported and furthered Klein's idea that there should be attached to the government-sanctioned society a scientific advisory board consisting of four professors at Göttingen: Klein, Prandtl, Runge, and Wiechert. Through Klein's leadership as spokesman for the advisory board, he was able to muster the financial resources for the society to construct Göttingen's first small wind-tunnel laboratory. The Göttingen Association and the Prussian Ministry of Education each contributed 10,000 marks. In the same year that construction of the laboratory began (1907), Prandtl received the promotion to (full) professor of applied mechanics.[15]

Development of the aerodynamical laboratories at Göttingen through World War I involved Klein as much as Prandtl, even if after 1910 Prandtl had received wide acclaim for his boundary-layer theory and wielded his own power in planning Göttingen's research.[16] At the instigation of Klein, Emil Wiechert, the professor of geophysics, received from the firm of Krupp in 1909 a grant of 10,000 marks per year for three years to develop an inherently stable, unmanned flying machine.[17] The airplane was to improve the quality of observations of electrical and meteorological conditions at high altitudes. Direct support of this research was not Krupp's only purpose. The firm also intended by it to induce other donors to join Klein's aeronautical enterprise.[18] Not surprisingly, Krupp immediately received an invitation to join the Göttingen Association, which he accepted as quickly.[19] Krupp's money was channeled through the association. That same year the Society for the Study of Motor Airships, beset by internal bickering, disbanded, and the wind-tunnel laboratory reverted to the university.[20] Klein outlined his ambitious plans for research there in a paper that he read before the Aeronautical Union of the State of Lower Saxony. In all, the sums allotted for aeronautical research totaled to a substantial 30,000 marks, which could, Klein assured his listeners, "be applied only for pure science."[21] This meant, in the context of the "Peace of Aachen," that engineering development was excluded. Pure science, however, could include the study of applications.

"Klein desired," noted Prandtl, "in the field of flight, as in everything, [to maintain] close contact between theory and practice; so he conceived a congress which brought together in Göttingen from 3 through 5 November 1911 the best-known representatives of aeronautical science and technology, and he directed that congress personally." The participants in the conference voted a mandate to the Göttingen Association to continue the exchange. From this resolution Böttinger took the opportunity to lead in founding a German Scientific Society for Aviation, which soon sponsored publication of the two-year-old but already important journal *Zeitschrift für Flugtechnik und Motorluftschiffahrt* (Journal for Flight Engineering and Airship Travel).[22]

In 1911 also, Prandtl applied to the newly founded Kaiser Wilhelm Society for the Advancement of Sciences for a grant to build a larger, permanent wind-tunnel laboratory, with the support and encouragement of Klein. Böttinger, who had become a "senator" to the Kaiser Wilhelm Society, sponsored the plan in the society and won approval for partial support in mid-1913.[23] Another year of lobbying in the Prussian ministries brought a promise that the state would provide the rest of the money. One-half of 200,000 marks for construction would come from the Kaiser Wilhelm Society, the other half from the Prussian Ministry of Culture.[24] But with the outbreak of war in August 1914, the plan languished in the ministry, victim of the policy that all financial resources of the state should be devoted to the conflict. It was expected that after a quick victory such plans could be reconsidered. As the war continued, Klein, Böttinger, and Prandtl interested the Ministry of War and the Naval Office together in May 1915 in funding the new aerodynamics institute.[25] Despite a move by several of Klein's long-standing opponents in engineering to undermine the plans for academic aeronautical research at Göttingen, military support materialized.[26] Within the month the two offices, prodded by a warm recommendation from aviation enthusiast Prince Heinrich of Prussia, awarded money for construction and additional sums for operation during the war, 300,000 marks all told.[27]

In the new wind-tunnel laboratory tests were begun in Spring of 1917.[28] Though Klein had been in retirement since late 1912 because of a mental breakdown that prevented him from bearing official responsibilities, he could not relent in his efforts to build the centers of applied science at Göttingen. Again, with his ally Böttinger, he had used his powers of persuasion, which some of his contemporaries marveled were "godlike."[29] Prandtl, too, observed this (in his obituary of Klein): "his power over men was astounding."

In another way Klein's work at Göttingen influenced the course of early aeronautical science and technology. As organizer and editor of the monumental *Encyclopedia of Mathematical Sciences Including Their Applications,* he compiled a collection of definitive studies that became the standard reference in mathematical physics. Early in the thirty-year enterprise Klein solicited the esteemed S. Finsterwalder, professor of mathematics at the Munich polytechnic (and, incidentally, one of Prandtl's teachers), to write an essay on aerodynamics.[30] This review article is significant in the history of aerodynamics because of its comprehensive scope and because it was submitted in August 1902. The date was more than a year before the Wrights achieved their powered flights at Kitty Hawk, North Carolina, and two years before Prandtl introduced his theory of the boundary layer. It is therefore a kind of prenatal record of the theoretical science we now call aerodynamics. More to the point, however, it was then a rare compendious account of the state of the art of aerodynamics, a first reference to be found in much subsequent research in the field. Klein's encyclopedia as a whole, moreover, provided the model for the later publication of *Aerodynamic Theory*, the six-volume encyclopedia of the science of flight that William F. Durand edited in the mid-1930s, with consultation of Prandtl and several other aerodynamicists, under a grant from the Guggenheim Fund.[31] Durand's encyclopedia made available much of aerodynamics to America. It remained a useful reference in aerodynamics more than thirty years after it was published.

Ironically, at about the time Kármán was coming to America, one leader of German aerodynamics lamented the lack of acceptance of the applied mathematical sciences in Germany. Richard von Mises of Berlin, introducing a visiting lecturer, the applied mechanics professor from Bulgaria K. Popoff, observed in May 1926: "To be sure, discerning men of broad vision like Felix Klein in Göttingen and director of the Prussian system of higher education Althoff sought several decades ago, with energy and understanding, to reform [the teaching of mathematical physics], but sadly it must be said that they did not reach beyond tentative victories."[32] In touch with other centers of applied mathematics and mechanics as a leader in the field and editor of its major journal, Mises was qualified to draw such a conclusion. At that moment it might have seemed to a neutral observer that Klein's efforts would have more impact in America than in Germany. Max Born had written a similar lament to Kármán in late 1922, and in 1926 Millikan and Guggenheim were working to strengthen aerodynamics in the United States. Far from neutral observations, however, Born's and Mises's regrets were a sign of the Weimar period in German history. In this setting, the beleagered physicists and mathematicians felt persecuted, perceiving both vilification and an actual decline of scientific and technical study. Klein's efforts would bear major consequences later, in the Germany of Hitler and in America at centers like Caltech.

In this one instance at least, the laboratories of applied

mechanics, aerodynamics, and hydrodynamics belied Mises's pessimism. Prandtl's facilities grew steadily (but for a temporary reduction in the work force and resources in 1919). After completion of the first wind tunnel, plans to build a larger wind-tunnel facility, which came to fruition as the Kaiser Wilhelm Institute in 1917, followed almost immediately. A second, hydrodynamics research laboratory opened in 1925, thus bringing to completion, just at the crescendo of lament for the death of applied science, a large applied research complex in the elite German university. Even in a time of economic need in Germany and of great animus directed against Germany among scientists of the allied countries, the quality and worth of Prandtl's work was granted internationally. Someone touring the scientific centers of flight in Europe could hardly fail to be directed to, and upon close inspection to be impressed by, the applied physics, particularly the aerodynamics and hydrodynamics laboratories of Göttingen.

The greatest benefits of Klein's fund-raising and organizing efforts in aeronautics had accrued to Ludwig Prandtl, director of aeronautical research at Göttingen. Prandtl was said, on occasion, to be naive in matters of social or political sensitivity[33]; so it may have been fortunate that he had Klein for an advocate through the war. Prandtl did succeed in converting an offer to replace his teacher in Munich in 1923 into the opportunity to build a new hydrodynamics institute at Göttingen, but his excellent academic reputation minimized the need for political astuteness in this negotiation. On aeronautics' and Prandtl's behalf, Klein was ever at pains to advance the cause of building or expanding institutes of research, beginning in 1906 with plans for the original "model testing institute," then from 1910 through 1916 for general research support and the larger "aerodynamic testing institute."

5

THE ACHIEVEMENT OF LUDWIG PRANDTL

Prandtl's research, even before he came to Göttingen in 1904, was the embodiment of Klein's ideal of binding together theory and practice; as such it enjoyed Klein's full support. In the words of Theodore von Kármán on the occasion of Prandtl's fiftieth birthday, in 1925, "Perhaps what one must wonder at most in Prandtl's scientific method, the direct connection of general, abstract theorems with experimental facts and practical applications, is pure, unadulterated Göttinger Tradition, which, adapted by F. Klein in new form and to the demands of the technical century, has undergone a rejuvenation."[1] Kármán's pronouncement not only attaches Prandtl to Klein's movement of reforming technical research, but it reveals in its tone the high regard in which Kármán expected his audience (of aeronautical engineers) to hold that reform and those associated with it. Klein was the deliverer of applied science, and Prandtl was his disciple.

In depicting the state of the science of fluid flow at the beginning of the twentieth century, Hunter Rouse, the civil engineer and historian of hydraulics, has described how Prandtl's work fulfilled Klein's ideal:

Civil engineers had long since developed an essentially empirical sort of hydraulics, as exemplified by the many tabulations of coefficients in the textbooks. . . . At the other end of the spectrum, mathematical physicists had refined a subject known as classical hydrodynamics, best exemplified by Horace Lamb's *Hydrodynamics*, first published (in Great Britain) in 1879 under the

title *A Treatise on the Mathematical Theory of the Motion of Fluids.* . . .
Neither subject, however was much use in the new profession of
aeronautics. On the one hand, the theoretical approach had little
direct contact with reality; on the other hand, pure empiricism
did not lend itself to application beyond the conditions of observation. What was needed was a combination of the good points of
each subject, without its weaknesses. Just such a hybrid approach
was provided early in the century by Ludwig Prandtl of Göttingen, Germany, and his many followers, nearly all of whom
were mechanical engineers. Proceeding from Prandtl's boundary-layer hypothesis of 1904 (that a state of flow can be approxi-
mated by a wall zone of viscous influence and an outer zone of
irrotational motion), these men developed a quasi-analytical
approach now known as fluid mechanics, in which the mathematics was simplified to the greatest possible extent still in accord
with experimental indications.[2]

Rouse emphasized that Prandtl's 1904 paper introducing the
boundary-layer hypothesis[3] was one of about eighty read before
the Third International Congress of Mathematicians at Heidelberg and "apparently received only passing attention from the
mathematicians who heard it."[4] But its significance was not lost
on Felix Klein, who also attended. With extension and elabora-
tion by Prandtl and his students, and with its confirmation
through the use of experimental resources that Klein put at his
disposal, the central idea of the boundary layer would take its
place in the small core of fundamental concepts constituting the
theory of aerodynamics.

In several ways, this early paper typified Prandtl's future mode
of research. It set out from very general introductory remarks to
arrive at precise statements about the flow of fluids around certain
obstacles of different shapes. The arguments were essentially
physical, based upon Prandtl's intuition reinforced by his
confirming experiments. The paper was devoted about equally to
explaining the theory and to verifying it with drawings and
photographs of flow, taken from his own calculations and experiments in a water canal. It presented an approximate solution to

the problem he posed, arrived at by numerical computation. In this respect and in others the paper was quite different from a typical work of Theodore von Kármán's. Many of his subsequent works bore similar attributes. The paper is easily surveyed and deserves our attention for what it reveals about Prandtl's scientific turn of mind, which was a part of what appealed to Harry Guggenheim's advisors and others of the Americans who visited Germany in the first quarter of this century.

Prandtl posed the problem he was to solve by noting that classical hydrodynamics had usually treated the flow of inviscid fluids. If viscosity has been introduced in two- and three-dimensional problems, then the inertia of the fluid had been neglected, in order to arrive at any solution at all. The reason lay in the "unpleasant properties" of the Navier–Stokes differential equation that governed the motion of the fluid: integrals of the equation were in general impossible to find. But if the motion of the fluid was slow enough, its inertia could be neglected and the problem would be simplified. Or, on the other hand, if the motion was fast enough, inertial effects would far surpass those of viscous resistance. In technological (nautical and hydraulic) problems, said Prandtl, the second situation almost always obtained. Nevertheless, one knew that solutions of the resulting equation were often in poor agreement with experiment. Thus, more properly, viscous, or internal, resistance had somehow to be considered along with inertia.

Prandtl thereby came to the central idea of this paper: "I have set myself the task of investigating systematically the motion of a fluid of which the internal resistance can be assumed very small. In fact, the resistance is supposed to be so small that it can be neglected wherever great velocity differences or cumulative effects of the resistance do not exist."[5] Prandtl's "systematic investigation," in this paper at least, consisted of treating no more than a few "single questions" in broad outline. In one-half page Prandtl produced several simple results from general consideration of the

Navier–Stokes equation, results whose technical and unoriginal nature render their derivation inessential here.

The most important point of the paper was the more physically intuitive, and thus the more accessible to his listeners, as it is to us: "By far the most important question of this problem is the behavior of the fluid at the surfaces of the solid body." Motivated by two insights, that the fluid must be immobile at the surface of the body and that the fluid must follow classical hydrodynamic streamlines only a short distance from the surface, Prandtl postulated the existence of a thin "transition layer," the only region where viscous effects are evident:

The physical processes in the boundary layer between fluid and solid body are sufficiently taken into account if one assumes that the fluid adheres to the surfaces; that therefore the velocity everywhere along them is equal to zero or to the velocity of the body. If now the viscosity is very small and the path of the fluid along the surface is not too long, then only a short distance from the surface the velocity will have its standard [free-stream] value. In the narrow transition layer then, the great velocity variations, despite small coefficients of viscosity, give rise to noticeable effects.[6]

It is evident in what followed that Prandtl had already decided this boundary layer would merit further investigation.

He laid out a plan for solving boundary-layer problems by suggesting that such problems are more tractable if one systematically neglects certain variables in the general differential equation. If the viscosity is small in the second order, then the thickness of the boundary layer will be small in the first order, as will be all components of velocity normal to the free-stream flow. "The transverse differences of pressure can be neglected, as can any curvature of the streamlines." These "systematic omissions" may simplify the Navier–Stokes equation to a degree where it is soluble.

Prandtl wrote the governing equation that arises from applying his simplifications to the problem of flow in two dimensions, the only one he had yet attacked. One could solve problems of this form (for the velocity of the fluid as a function of its position) through a step-by-step calculation with help of one of the known methods of approximation like that developed by his German colleague M. Wilhelm Kutta,[7] whom he cited. In the simplest case of a fluid flowing uniformly along a flat, thin plate—a problem that A. Gustave Eiffel, Albert Francis Zahm, and other early aerodynamicists had studied experimentally[8]—he presented his results for the resistance of the plate to the flow and for the state of motion, but did not demonstrate his path to them. He did not reduce his formula for resistance to the nondimensional parameter called Reynolds number but, rather, expressed it in terms of measurable quantities of plate width, plate length, velocity of undisturbed fluid, density, and viscosity. The state of motion he expressed by a plot of the transverse velocity u along the plate as a function of transverse and normal coordinates x and y (figure 5). For applications, "the most important result of these investigations" was an evident separation of flow from the surface of the plate. Hermann von Helmholtz had noted this separation in inviscid fluids, but Prandtl now showed that the analogue to Helmholtz's "separation layer" was a region of viscous fluid in rotation, arising from friction between the fluid and the surface, a so-called vortex layer or vortex sheet. As the viscosity approaches zero, only the thickness of the vortex sheet diminishes; the same flow diagram is retained as in Helmholtz's theory.[9]

Prandtl stated the consequence of his hypothesis of the boundary layer: "According to the above, the treatment of a certain process of flow therefore reduces to two parts in mutual interaction: one has on the one hand a *free fluid*, which can be treated as inviscid according to the vortex principles of Helmholtz, on the other hand the transition layers at the fixed boundaries, whose motion is governed by the free fluid, but which for their part give the free motion its characteristic features by the emission of vortex layers."[10]

Partly because the paper was written for oral presentation, but also because it inevitably reflected Prandtl's method of comparing theory to experiment, he included several drawings of the streamlines of flow around different obstacles, which showed clearly the phenomena of separation and vortex formation (figure 6a). As a crowning touch, Prandtl presented photographs of experiments he had carried out, some with configurations whose solutions he had already calculated (figure 6b). He described the evident flow of water around the obstacles, pointing out where separation occurred and explaining why.

The truly unaeronautical nature of this study deserves emphasis. Only later, in 1906 so far as sources at Göttingen indicate,[11] did Prandtl become interested in the theory of flight. Then the work became of great use in calculating frictional drag of surfaces in an airstream. Indeed, rather than his interest in flight leading Prandtl to the discovery of the boundary layer, cause and effect were likely reversed. The study, and Klein's interest in aviation, spurred Prandtl to consider the hydrodynamics of flight, which he had not treated, though others like Kutta and Joukowski had, prior to his coming to Göttingen.

The paper shows a definite scientific astuteness. Prandtl recognized that in such a forum as the 1904 Heidelberg congress he would have to capture the attention of his audience quickly and hold them through the main points of his analysis. Here was no place for detail. Much of what the mathematicians might elsewhere construc as important he excluded; for example, the specific method of solution is nowhere to be found in the paper. His essential innovation is the physical hypothesis of a boundary layer. He presented his result in graphical form, but it is not clear how he really solved the problem, except that it was by numerical calculation. It is clear from his graphs, a formula, and the photographs of his water-canal experiments that his idea succeeded. The work, which was by no means his first, thus suggests a greater perspicacity in Prandtl as scientific salesman than some commentators have granted him. The paper was a meager eight pages long, its ideas the foundation of the modern fluid mechanical theory of drag.

Theodore von Kármán in his autobiography projected airplanes into Prandtl's hydrodynamic investigation. Nevertheless, his statement is a clear and proper summary of the essence of the work; and of note since it was against Prandtl's achievements that his own work would be compared:

He ingeniously assumed that the total effect of friction on any part of the airplane can be estimated by the trick of restricting the investigation of the forces to a thin sheet of air close to the surface, which he called the "boundary layer." The rest of the surrounding atmosphere was not affected by friction of the moving plane, and its motion could still be explained by the air circulation theory. Prandtl's simple concept of the boundary later marked one of the most important advances . . . in the theory of flight.[12]

Prandtl's methods were a mix of pragmatic approximation and elegant analysis, of quasi-theoretical considerations and careful plots and experiments. Such a combination of talents was unusual in German engineering research. Although they were fully in line with Felix Klein's ideology of practical mathematics, it was in fact August Föppl, Prandtl's teacher at Munich, and not Klein, who early had counseled Prandtl in those methods. The historian of physics Gerald Holton, in arguing that the treatise of this same Föppl titled *Introduction to Maxwell's Theory of Electricity* was a major influence on Einstein's invention of the special theory of relativity, has pointed out that Föppl had a "special talent as a teacher and writer," a certain pedagogical gift that brought him a respectful following.[13] Although Föppl's students had sometimes complained that his lectures proceeded too slowly, he found that this pace was necessary to lay the foundations of technical physics properly. These he did, going back to first principles. So much Föppl claimed in the foreword to his immensely popular *Lectures on Technical Mechanics*,[14] which grew out of his teaching. Föppl, who became Prandtl's father-in-law in 1910, had advocated in his lectures understanding the physical laws that underlie the rules of engineering design. Kármán recalled, "Prandtl was affected by this marriage of theory and design [of Föppl's] and transmitted some of the philosophy to me, which firmed up my own thinking."[15] Thus, Felix Klein brought to Göttingen in the person of Ludwig Prandtl the legacy of Föppl's general method that served to invigorate the university in the technical field of aerodynamics.

Prandtl's research and his importance to physical science and to Göttingen's program of applied physics exceed the bounds of aeronautics. In combining fundamental physical arguments with specific applications, he contributed to the fields of hydraulics, hydrodynamics, aerodynamics, solid mechanics, and heat transfer. In the three fields of fluid flow his works were seminal. Since our primary interest lies in aerodynamics, I shall pass over the rest and but mention his major contributions in that field before and during the time when Guggenheim and Millikan agreed to seek a European scholar of high accomplishment to lead the laboratory at Caltech.

In a Prandtl *Festschrift* that appeared in 1925 as a separate number of the *Zeitschrift für Flugtechnik und Motorluftschiffahrt*, Kármán surveyed Prandtl's works at age fifty.[16] This and the 1930 biographical sketch attached to Prandtl's citation for the second Daniel Guggenheim Medal[17] provide similar professional estimates of the significance of Prandtl's work and of its reception in the 1920s. Kármán noted that Prandtl's investigations of the boundary-layer theory had been extended by many researchers since 1904, on the one hand by improving methods of integrating the equation of motion, on the other hand by carrying out precise experimental researchers. But Prandtl's major accomplishment for the technology of flight was undoubtedly his wing theory, and in particular his discovery of the so-called induced drag of a wing. These ideas were first expressed in lectures in 1910, though they were not published in a full presentation until the two-part paper with the simple title "Wing Theory, I. and II. Communications" appeared in print at the end of the war.[18] Through his theory one could approach a rational, analytical model of the airplane, especially in predicting the influence of wing span, angle of incidence of the wing to the airflow, multiplicity of wings, and formation flying on the performance of aircraft. Kármán placed Prandtl's wing theory in the context of what was known before and evaluated its significance:

The analogy between an aircraft wing and a vortex filament [*Wirbelfaden*] was indeed known; also the fact that with finite wing span the circulation at the wing tips cannot simply cease, but must release so-called "vortex trails" [*Wirbelzöpfe*] was stated by Lanchester; Prandtl recognized with gifted insight, however, that the consequent carrying out of the Helmholtz vortex theory under the assumption of lightly loaded wing surfaces must lead to a complete theory of ideal wings, which supplies all information on the distribution of lift, performance requirements, etc., that are independent of the profile resistance. To the discovery of the minimal resistance of the ideal wing, the so-called "induced drag," belongs approximately the same significance for aircraft production as the discovery of the Carnot Process for the construction of heat engines: both provide a yardstick, by which the quality of construction is measured, a principle of order, through which otherwise unclear experimental material becomes in one stroke clear and understandable.

But what perhaps is most important is this: that there is today hardly a plant in the world where aircraft are designed and the idea of induced drag is not considered.[19]

To a technical audience, this enthusiastic endorsement would have carried considerable force in its comparison of Prandtl's concept of induced drag to the Carnot process. For out of Sadi Carnot's conception of a century earlier, by which one grasped for the first time how most efficiently to convert heat to work, came not only a powerful tool for engine design but the very foundation of the theory of thermodynamics, for which both physical scientists and engineers had great reverence. If Prandtl's principle proved as important to aerodynamics in this century as Carnot's was to heat theory in the nineteenth century, his would be a high elevation indeed.

Aside from Prandtl's own superb research, two circumstances combined to further the quality and amount of work conducted under his direction at Göttingen. The first was the impact of the war, which, through the labors of Klein, aided Prandtl's fortunes in acquiring a wind-tunnel institute much larger than the one he had directed before the war. With the financial support of the War Ministry, Prandtl had acquired by late 1918 a staff of fifty,[20] and evidently wartime financing had enabled much of his wing theory to be tested.[21] The second circumstance was that Prandtl's groundbreaking work and large laboratory facilities gained him renown that attracted scores of high-quality graduate-research students. Hunter Rouse cites, among the most famous of these, Theodore von Kármán; Albert Betz, Prandtl's principal assistant; Walther Tollmien, theorist and successor to Prandtl in the Göttingen chair; Ludwig Schiller, who worked in hydraulics; Jakob Ackeret, contributor to the theory of supersonic flow; Oskar Tietjens, who published Prandtl's lectures; Hermann Schlichting, a further developer of the theory of the boundary layer; Carl Wieselsberger, who succeeded Kármán at Aachen in 1934; Otto Flachsbart, who studied phenomena of drag on buildings; and Johann Nikuradse, who studied problems of flow resistance in pipes.[22] He fails to mention Max Munk, who contributed notably to the theory of induced drag and who for a time in the early 1920s was the chief theorist of the U. S. National Advisory Committee for Aeronautics. Although Prandtl's lecturing could not compare with the luminary David Hilbert's[23] he gathered this loyal following of students over more than four decades of active research. In 1925 he had reached a new level of eminence, having undertaken to direct his second, hydrodynamics institute.[24] He was the acknowledged, if not unchallenged, leader of German aerodynamics and fluid mechanics.

6

KÁRMÁN AT GÖTTINGEN AND AACHEN

In truth there was but one challenger to Prandtl's preeminence in German aerodynamics in 1926: Theodore von Kármán, former student of Prandtl's and Director of the Aachen Aerodynamics Institute. The son of a well-known professor and educational reformer at Budapest, Kármán (figure 7) had earned his first degree with honors in 1902 from the Budapest Royal Polytechnic Institute ("The Royal Joseph") in mechanical engineering.[1] In school he had felt a special attraction to scientific theory, as he later recalled, which would draw him to Prandtl's institute. Kármán spent three years after the polytechnic as a research engineer in Budapest, simultaneously assisting the professor of mechanical engineering at his alma mater and consulting for industry. With a theoretical bent and demonstrated excellence, however, he applied to and received a fellowship in 1906 from the Hungarian Academy of Sciences to study at the University of Göttingen. At Prandtl's institute of applied mechanics Kármán received his doctorate after writing a dissertation on inelastic buckling that was soon recognized as a significant contribution to the mechanics of solids. Only after he had been at Göttingen for two years was his professional interest excited in the theoretical aerodynamics that Prandtl had initiated. From 1908 to 1912 he was a research assistant and nonfaculty teacher (*Privatodozent*) at Prandtl's laboratory; in late 1912 he taught applied mechanics at the mining academy in Selmeczbanya, Hungary; and then in 1913 he sought and received a call to the Polytechnic Institute at Aachen, where he became professor of mechanics and technical aerodynamics and director of the new Aerodynamics Institute.[2] At approximately the same age, thirty-two, Prandtl had received his first directorship of an institute, but 1913 was a decade after the Wrights' successes with powered flight and well into the age of aviation. So it was that Kármán began his independent career, educated in the values and familiar with the methods of aerodynamics research, as Prandtl had not been in 1904. If it were not for the interlude of the First World War, he might have challenged his professor well before the mid-1920s, for the works of Kármán and Prandtl, though sometimes different in method, began early to follow parallel paths in developing a theory of the airplane.

Kármán's and Prandtl's modes of thinking can be compared conveniently thanks to the existence of a paper on aerodynamics that is analogous, in the role it played in Kármán's career and in its intrinsic significance, to Prandtl's paper on the theory of the boundary layer. This work, "On the Mechanism of the Resistance That a Moving Body Experiences in a Fluid," was presented in late 1911 to the Göttingen Scientific Society.[3] It represented Kármán's first substantial work in aerodynamics and the beginning of a line of research on vortices and turbulence that extended over most of his career. In this paper he introduced the detailed mathematical analysis of the stability of two parallel rows of vortices known now as the "Kármán vortex street" though it had been observed earlier.[4] I want to examine the work in some detail, even if it requires a little effort, for it suggests a good deal of his early style and method and thus in part why the principals at Caltech recruited Kármán for their institute.

His treatment of this problem succeeded in deriving by mathematical analysis a quantitative, experimentally testable result about vortex behavior that was *not* known before and which could readily apply to many technical problems. Kármán's point of departure lay in his discussion of the difficulties of solving certain hydrodynamics problems at different viscosities, from zero to infinity. In this respect his paper began like Prandtl's 1904 work; but it soon diverged.

The thread of argument in Kármán, like that in Prandtl, was distinct and taut. He stated a general formula for the resistance of a solid body in an incompressible fluid that Osborne Reynolds had written first in 1883.[5] It was the simple relation that resistance W of a body in a fluid stream is equal to the product of viscosity and velocity of the fluid μ and U, of a characteristic dimension of length of the body l, and of an undetermined function f of the nondimensional parameter called Reynolds number $U l \rho/\mu$ (where ρ is the fluid density). This function depends on the shape of the body and the choice of l. British physicist George G. Stokes had investigated the limiting case of this law where Reynolds number is very small, that is, at high viscosity or low velocity of the fluid. In the other limit where the Reynolds number goes to infinity, however, problems of flow had proved more difficult to treat. Yet they were "no less important, because almost all technically important cases of liquid resistance and in particular the questions of air resistance belong to this [case]."[6] Here Kármán's remarks were reminiscent of Prandtl's. But Kármán acknowledged, as Prandtl had not in his 1904 paper, that he had in mind aircraft and propellers. Their moderate speed would allow him to treat air flowing past them as an incompressible fluid, and so allow the use of Reynolds' formula for resistance. The limit of infinite Reynolds number corresponds to inviscid fluids, and this was the limit he would investigate. The problem's essence did not lie in the viscosity of the fluid, as it did in Prandtl's problem. Thus the methods of classical hydrodynamics would suffice.

Kármán pointed out, nevertheless, that neither the hypothesis of continuous flow past the obstacle, as adopted first by Jean d'Alembert in the mid-eighteenth century, nor the more recent theory of "potential motion" of Helmholtz, Kirchhoff, and

Rayleigh—in which it is assumed that a surface of discontinuity forms at the edges of the obstacle, with dead water behind it—leads to a proper solution.[7] The former was physically useless, since it implied zero resistance, a well-known difficulty referred to as d'Alembert's paradox; the latter would lead to a functionally correct form for the resistance, which, however, deviated in value from that measured. The discontinuous theory failed, said Kármán, because it inadequately treated the "suction effect" behind the obstacle, which had recently been observed as the major contributor to resistance. (Note that Kármán acknowledged recent experiment as the source of one aspect of his theoretical insight into this problem.) Also, the discontinuous theory assumed that the flow is stationary with respect to a coordinate system in the rest frame of the obstacle. This presumption, Kármán noted, "corresponds in no way to reality"[8]: there is turbulence trailing away from the obstacle rather than the "dead water" extending to infinity behind and accompanying the obstacle that the Helmholtz theory postulates.

As fluid flows around an obstacle, a separation of the flow occurs that, when the flow reunites behind the obstacle, leads in general to an instability of the "surface of separation" between the two flows. This surface, which Helmholtz had designated a "vortex sheet" and which had been investigated in depth by William Thomson,[9] showed a marked tendency in experiments to roll up around certain lines normal to the plane of flow,[10] until eventually one found a series of isolated vortex filaments in or near the plane of the vortex sheet. These, Kármán knew from observing his young colleague Hiemenz's experiments in Göttingen,[11] would align themselves in the simplest stable configuration along two parallel rows behind the obstacle (figure 8). This qualitative discussion and references to the literature constituted his only attempt to explain the *existence* of vortices behind the obstacle; he made no explicit reference to viscosity or the boundary layer, nor did he even pose the general, complicated problem of flow. His interest lay rather in this: *Assuming* the generation of two parallel rows of vortices trailing behind the obstacle, what is the character of the motion in steady state?

Kármán accomplished two mathematical analyses in this paper. First he showed that given two parallel rows of vortex filaments there is but one stable configuration, in which vortices of opposite rotation alternate above and below, in a staggered configuration, at fixed horizontal separation and height. He had claimed without argument that only two stable configurations were conceivable: one with vortices directly opposite each other and one with vortices staggered. Since Kármán recognized in advance that the problem was to find conditions for stability of a certain vortex configuration, he proceeded by calculating deviations from the equilibrium position and then showing that there was a condition for which the deviations would remain small. This condition could be fulfilled only in the second configuration. The stable vortex arrangement occurred when the ratio of height to horizontal separation was equal to (Arccosh $\sqrt{3}$)/π.[12] He illustrated this result in a simple drawing of the "vortex street" (figure 9). Kármán was then able to calculate the constant velocity of the quasi-steady motion of the vortex street with respect to the obstacle.

Second, he showed that the mechanism of resistance is a transfer of momentum from the obstacle to the trailing vortices. He calculated this transfer, and from it produced a relation between the resistance and the ratio of height to horizontal separation between vortices. The relation provided an empirical test of the theory, though he cited no data.[13]

After the first analysis, Kármán had tied up some loose ends of the argument. At first he had solved the problem of determining the condition of stability by calculating contributions to the velocity of one free vortex from two rows of *fixed* vortices, and this he now acknowledged was an imperfect picture. But, he asserted, an analysis treating freely moving vortices would lead to the same condition of stability. As he admitted, Kármán left unsolved the other part of the mathematical problem, namely, to prove that other arrangements of parallel lines of vortices strive toward the single stable configuration. He did elicit a plausible "physical" argument in favor of the proposition.[14]

Three remarks concluded Kármán's discussion. First, there was an apparent contradiction in all this, in that a theory of inviscid fluids had been used to explain the position of vortex filaments, which, according to the Helmholtz theory, could not originate in such a fluid. One might dispel any doubts of the validity of the analysis by recognizing that the mechanics of ideal fluids is the limiting case of mechanics of viscous fluids at large distances from the obstacle. Prandtl's theory of the boundary layer had perhaps first brought this thought to a clear expression, Kármán wrote. Even though the forces of friction go to zero in the limit of zero viscosity, the magnitudes of the vortices, which arise from friction, remain finite in the same limit. And it is from these vortex magnitudes that the resistance is determined. Although Prandtl's theory was not a necessary part of Kármán's discussion of vortex magnitudes, the boundary-layer theory accounts for the existence of the vortices.

Second, it might seem curious that his solution for the motion was an "unsteady picture" (in its alternate generation of vortices). He accounted for this by noting that the unsteadiness arose from his treating the ideal, limiting case of an infinitely extended body. The oscillation would "most probably" disappear if one solved a real problem in three dimensions, such as that of flow past a (finite) rotating body. This problem might yield a stable, steady configuration of helical vortex elements. The three-dimensional case presented certain greater difficulties in calculation, however, and would have to wait for further study.[15]

Third, and perhaps most interesting for the historian of science, Kármán pointed to the analogy between the interaction of vortices and electromagnetic forces, which Helmholtz had first noticed. According to the analogy, the velocity field of the vortex filaments corresponds to the magnetic field from a linear conductor, if one substitutes current for magnitude of the vortex and the magnetic force for the velocity into the hydrodynamic equations. "The existence of vortex filaments at the surface of the obstacle would then correspond to current supply and the vortex filaments that follow slower behind the body may be represented by light linear conductors that free themselves from the body and determine their own shape and motion by their magnetic field."[16] Just as Kármán likened Prandtl's wing theory in 1925 to the Carnot process, here he called upon his knowledge of electrodynamics for an illuminating analogy in engineering science. Another engineer might have ignored the relation, if he noticed it at all, but Kármán, a follower of both the history and practice of theoretical physics, could hardly miss it. If he knew well the theory of electromagnetism, then the analogy would have served him with a familiar means to visualize vortex interactions.[17] Kármán valued the founding studies of others in his field as well as the heuristic power of borrowing from other fields. Not only did he cite the studies in hydrodynamics of Helmholtz, Kirchhoff, and Rayleigh in his introduction, but he had referred to William Thomson's original paper on the generation of vortex sheets as it had been republished, in English, in his *Mathematical and Physical Papers*. The work surely took some effort for Kármán to assimilate.[18]

Kármán cited one other theoretical source—the ubiquitous *Hydrodynamics* of Horace Lamb in its second edition of 1895. From Lamb's discussion of the laws of motion of vortices, Kármán derived a quantitative expression for momentum transfer by vortex shedding, the foundation of his analysis of the mechanism of resistance to flow. The thread of analysis passed from Lamb to Kármán and back to Lamb again. For Lamb found Kármán's work, as developed later and more generally, to be sufficiently cogent to discuss at length in his sixth edition of the same volume, which was published in 1932.[19] By contrast, Prandtl had acknowledged only two sources of inspiration: Helmholtz, to whom he had to bow for founding the vortex theory and discovering the separation layer, and Rayleigh, who solely had attempted to solve the problem of viscous flow exactly, for one-dimensional problems.[20] Neither had contributed directly to Prandtl's solution of the boundary-layer problem. That discovery was grounded instead in Prandtl's physical reasoning, and in it he applied no preexisting analytical apparatus like Lamb's, as did Kármán.

The paper I have surveyed was the first and the least useful of three papers Kármán published over a year's time on the vortex street. Before the argument would become compelling, indeed even correct, Kármán had to relax the constraint that only a single vortex pair within the parallel planes of vortices be allowed to move. Upon concluding the more general calculation, Kármán found that the ratio of height to horizontal length was not (Arccosh $\sqrt{3}$)/π but (Arccosh $\sqrt{2}$)/π. He found also that his formula for the resistance of the obstacle had to be altered when he calculated the momentum transfer, because he had used too large a value for the momentum carried by a vortex filament, which he had taken from Helmholtz. Thus, while Kármán's first calculation had actually contained an erroneous physical assumption, his corrections brought the results into agreement with experiments on a cylinder and on a plate perpendicular to the flow, his own and others' experiments that he now reported. These results his student H. Rubach and he presented in a note published in the pure physics journal *Physikalische Zeitschrift*,[21] another version of which appeared under Kármán's name only in the *Göttinger Nachrichten*, as part two of the original paper.[22] The latter studies were the ones that Lamb abstracted in his text, and they came to bear special significance in the theories of hydraulics and flight in ensuing decades, because of their many uses.

In 1954, Kármán listed a few phenomena that his theory had explained over the years:

A French naval engineer told me of a case where the periscope of a submarine was completely useless at speeds over seven knots under water, because the rod of the periscope produced periodic vortices whose frequency at a certain speed was in resonance with the natural vibration of the rod. Radio towers have shown resonant oscillations in natural wind. The galloping motion of power lines also has some connection with the shedding of vortices. The collapse of the bridge over the Tacoma Narrows [figure 10] was also caused by resonance due to periodic vortices.[23]

Kármán declared in the same place, with evident delight at the many analyses that had come from his first little study, "I am always prepared to be held responsible for some other mischief that the Kármán vortices have caused!" But even earlier, before so many consequences of his principle of vortex shedding emerged, Kármán received recognition for the elegance and significance of the work. He recalled to his autobiographical collaborator, "The two papers published on the mechanism of the fluid and air resistance in 1911 made my name really internationally known."[24] The position in aerodynamics at Aachen followed a year later.

Kármán admitted that when he concluded the first calculation he consulted Prandtl about its publishability and received the answer, "You have something. Write it up and I will present your paper to the Academy."[25] Kármán was still a *Privatdozent* connected to Prandtl's institute and possibly uncertain of the reception, if not the value, of his work. He was, to be sure, an experienced young researcher, having written some fifteen scientific papers. But this was his first in aerodynamics, and he would want the imprimatur of the czar of aerodynamics. In any event he had best show his work to his boss.

Slowly their relationship changed, as the relative differences in age and experience diminished, until in the early 1920s, the evidence reveals, Kármán knew that he deserved to stand as Prandtl's equal. Others would have agreed, for the quality of his work and the excellence of his reputation both were evident to the world of applied science when Millikan considered hiring him in 1926. I shall but mention the major achievements that brought him this reputation, works he accomplished through the mid-1920s as director of the Aerodynamics Institute at Aachen.

Kármán succeeded Hans Reissner in 1913 at the Aerodynamics Institute, which had been built in 1911, and he designed a new Göttingen-style closed-circuit wind tunnel soon after he arrived (figures 11 and 12).[26] He devoted much of his energy before the First World War to developing this new center, but Hungary activated his army commission in 1914, thereby compelling him to direct research for the Austro-Hungarian Air Corps during the war. He could not resume his research at Aachen until 1919. Evidently it took two more years for him to set his institute on course, for he published only four papers between 1913 and 1921, and only two of them, written in 1914 and 1918 in collaboration with E. Trefftz, addressed problems of aerodynamics. The first dealt with longitudinal stability and longitudinal vibrations of aircraft.[27] The second was a modification of the Joukowski theory of airfoils. This theory was important for enabling the aerodynamicist to transform mathematically a given wing profile, through the method of conformal mapping in complex analysis, into a curve approximating a circle. It thus simplified the determination of aerodynamic properities of the profile.[28]

The year 1921 saw a rich harvest of research in aerodynamics in both Göttingen and Aachen. Kármán published three papers, the most important of which, "On Laminar and Turbulent Friction," built on Prandtl's boundary-layer theory.[29] Prandtl himself published the first volume of what would be the widely acknowledged and immensely significant four-volume series with a German title translating to *Results of the Aerodynamics Research Institute at Göttingen*. Kármán, at age forty in that year and perhaps feeling a certain urge to show himself Prandtl's equal in management and science (we do not know how he reacted to Prandtl's ribbing him at reaching "the age of discretion"),[30] more the internationalist than his mentor, set about organizing with the mathematician Tullio Levi-Civita a conference for the following year in Innsbruck to bring together the diffusely scattered practitioners of applied mechanics. There he delivered a paper titled "On the Surface Friction of Fluids," which laid out a first approach to a full theory of turbulence.[31] Kármán observed that there was as yet no such theory, and without one "there is little hope of being able to fully comprehend the empirical material." He suggested that one might take a statistical approach, a guideline that he followed in subsequent works.[32]

Out of the Innsbruck conference he and several others organized the International Congresses of Applied Mechanics. The first of these convened in Delft in 1924 (figure 13) and the second in Zurich in 1926. At the first, Kármán delivered a paper on a rudimentary theory of turbulence, whose importance was quickly recognized.[33] That same year he introduced the idea of effective width of structures that are composed of sheet material covering longitudinal stiffeners, as in an aircraft wing or fuselage.[34] "The effective width," wrote Hugh L. Dryden in a memorial essay on Kármán, "is that width of the sheet material which can be regarded as carrying the same stress as the stiffener."[35] Kármán wrote several other papers on elasticity and solid mechanics, which need not be cited here. The year 1924 was exceedingly fertile for Kármán, as he and Theodore Bienen developed a theory of propellers as well.[36] In 1926 Kármán delivered a review of the theory of elastic-limit states of solids at the Second International Congress of Applied Mechanics,[37] but perhaps more important was his calculation in the same year of the pressure distribution of air around airships, in a paper that stemmed from consulting for the Airship Factory at Friedrichshafen.[38] In sum, by 1926 Kármán had established himself as a developer of cogent ideas in the aerodynamics of wings, propellers, and airships; in structures and the general theory of elasticity; and in the fundamental theories of drag and lift.

What is one to make of Kármán's work before 1926? His scientific methods and Prandtl's had in common certain features. Like other accomplished theorists, both grasped the heart of a problem. Prandtl recognized the need to forego an exact solution and to analyze arithmetically the problem of the boundary layer; Kármán saw that it was the conditions of stability of two rows of vortices, and not the full description of motion of those vortices, that was needed to explain the vortex street. Both Prandtl and Kármán sought to apply full intellectual analyses. That is to say, even if the theoretical solution could not be accomplished without some reliance upon intuitive assumptions, even if the final working out of the problem was experimental, still Kármán's and Prandtl's fundamental concerns lay in discovering the laws of nature that govern problems of flow. In Prandtl's paper of 1904 and in Kármán's of 1911 at least, and for much of their careers, their interests were essentially scientific. Both began their major works at the foundations of physical law, an approach that came partially from the teachings of Munich's August Föppl. This partly explained the remarkable coincidence in their introductory discussions of where the current theory of flow was valid and where it broke down.

Their scientific differences seem less pronounced than their similarities. Whereas Prandtl clearly relied upon experimental data to confirm his theoretical conclusions, and delighted in performing experiment, Kármán's overriding concern was a theory of fluid flow in the abstract, as well as the engineering problems of hydraulics. Kármán had commenced his education in the study of elasticity (at the same time Prandtl was interested in it) but moved soon to airships, not to the flow of liquids that began Prandtl's research on the boundary layer. It was Kármán's lust for theory rather than practice that brought him to a place to study aerodynamics. But his motivation lay also in zeal for aeronautics, inspired by the successes of early flying machines he witnessed in 1908.

Beyond this, Kármán showed certain qualities not so evident in Prandtl. Kármán valued mathematical reasoning more than did Prandtl, and he remarked that Prandtl was in fact not a particularly clever mathematician.[39] Kármán's own style was often analytically more elegant. His facility at drawing together threads of reasoning from different disciplines, the breadth of learning in history and pure science—these strengthened the logic of his arguments and gave him the aura of intellectual sophistication. By his organizing activities at the various congresses, beginning at Innsbruck in 1922, we gather suggestions of his international ambition in applied mechanics. Where Prandtl and his associates represented German aerodynamics as a nationalistic enterprise, Kármán preferred to associate himself with a pan-European movement to apply mechanics to airplanes. Kármán exhibited a quality of personal discipline that surely underlay much of his success. He knew well to grasp only at small parts of a science of flight at first, slowly to build up a full theory of the airplane by understanding its elements through rigorous analysis. And yet, finally, I should not suggest that Prandtl too did not appreciate the need for this stepwise approach, for there were comparable scientific successes in the parallel and rival developments of the two men's careers.

Kármán's early work was at once eclectic in drawing upon different branches of applied mechanics and thematically united in trying to understand how airplanes fly, and thus to improve them. Although there is some evidence that about 1920 Kármán had wavered between pure and applied science, considering perhaps to pursue the former if the opportunity presented itself,[40] his dedication to aeronautical, usually fluid mechanical, science was unswerving after 1921. He was by far the best and the most successful of Prandtl's students, and might, but for circumstances, have equaled in recognition and status his illustrious mentor by the end of the third decade of the century.

Kármán's work followed in the traditions of Göttingen. Well founded in fundamental methods of physics, he recognized the boundaries between pure and applied sciences but sought whenever possible to cross them.[41] This, to a lesser degree an attribute of Prandtl's too, attracted Millikan and especially his advisor, the mathematical physicist Epstein, to the German aerodynamics laboratories at Aachen and Göttingen. Kármán's alliance with industry, through consulting contracts, was equally in agreement with the ideals that Felix Klein had vitalized at the turn of the century, and Kármán explicitly associated himself with Klein and his tradition.[42] Furthermore, Kármán directed not just theoretical research, but experimental investigations of his and Prandtl's theories. As at Göttingen, so at Aachen experiment was highly valued—perhaps more so, since the facilities were far inferior.[43] Even in 1911, as his vintage paper indicates, Kármán had understood the need for experiment to support theory, though he emphasized the reverse in later papers for polemical reasons. True, fewer experimental results were reported from the smaller Aachen institute than from the Göttingen laboratory, and Kármán bore a reputation among his students for being something of an oaf in the laboratory[44] (as do most theoreticians); but we should not construe this as suggesting antipathy to experiment. He gratefully accepted the latest experimental results when Prandtl offered them before publication from the Göttingen lab,[45] and he noted in his last years that most of his theories had been created to explain *observed* phenomena rather than to predict new effects from fundamental considerations.[46] Aachen and Göttingen had more in common than the different structure and purpose of their institutions suggested. Millikan was interested in looking at both laboratories and their directors for all these reasons and further because of the visibility and good repute that German aerodynamics research in general enjoyed in America. That visibility had developed early in the history of aviation.

1 C. Felix Klein. Courtesy of the
Niedersächsiche Staats- und Universitätsbibliothek, Göttingen.

2 The seal of the Göttingen Association for the Advancement of Applied Physics and Mathematics.

3 Carl D. T. Runge in 1926. From the
Mises photo album, courtesy of Geoffrey S. S. Ludford.

4 Ludwig Prandtl. Courtesy of the
Niedersächsiche Staats- und Universitätsbibliothek, Göttingen.

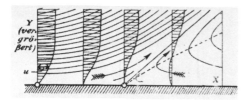

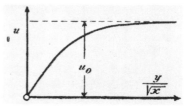

5 Prandtl's plot for flow over a flat, thin plate. From his paper, "Über Flüssigkeitsbewegung bei sehr kleiner Reibung."

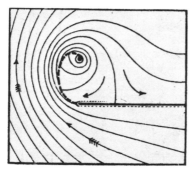

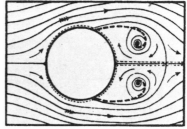

6 Calculated flow patterns (a) and photographic evidence of water-canal experiments (b) from Prandtl's 1904 paper on the boundary layer (opposite page).

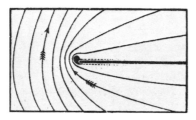

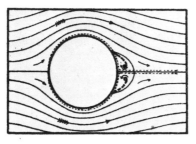

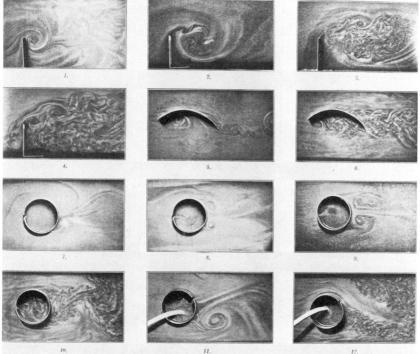

7 Theodore von Kármán, at far left, with his three brothers and sister Pipö, c. 1890. Courtesy of the Archives, California Institute of Technology.

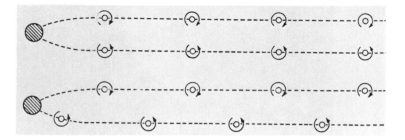

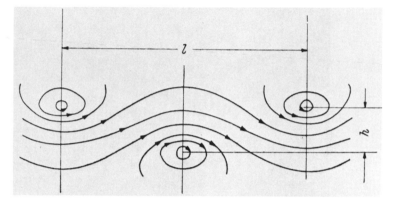

8 Double rows of vortices; symmetric
(upper) and asymmetric (lower) arrangements. Only the latter are stable.
From Kármán, *Aerodynamics*.

9 Kármán's drawing of three alternat-
ing vortices with characteristic dimensions of height *h* and horizontal separation *l* marked. From his paper,
"Über den Mechanismus des Widerstandes."

10 University of Washington model of
the Tacoma Narrows Bridge, 1941.
Note the wind-induced hump,
simulating conditions of failure on
November 7, 1940. Courtesy of the
Archives, California Institute of
Technology.

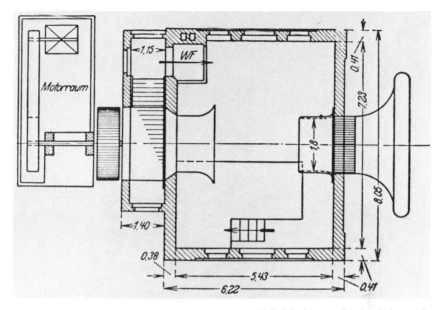

11 Original, open-circuit wind tunnel at Aachen, constructed 1912–13. Dimensions are given in meters.

12 Aerodynamics Institute at Aachen,
c. 1921. Note the ducting on the bal-
cony, which converted the old open
wind tunnel to a closed-circuit tunnel.
Courtesy of the Archives, California
Institute of Technology.

13 Participants at the First International Congress of Applied Mechanics, Delft, 1924. Kármán (fourth row, left of center) was co-organizer. From the Mises photo album, courtesy of Geoffrey S. S. Ludford.

7

AMERICAN RECOGNITION OF GERMAN AERODYNAMICS

About 1910 a seeming paradox existed in aeronautics: the country where powered flight of aircraft had first been achieved lacked a theoretical science of flight while the country in which that theory had been most highly developed was notably backward in building powered flying machines. German industry soon caught up in aviation, but in the United States programs of aeronautical research failed to develop until the mid-1920s. In the words of Millikan, the United States had through 1925 "lagged far behind Europe in the development of the science and art of aeronautics" since the work of the Wrights through 1908.[1] Yet the paradox was not complete, for America's advocates of aviation recognized the deficiency repeatedly in the teens and twenties and urged establishing institutions of aeronautical research. Their movement took the form of praise for the German style of research, praise not only for institutions (such as the laboratories at Göttingen and Berlin) but also for modes of thinking, especially of Ludwig Prandtl in Göttingen.

Although experimental studies in aeronautics in America had proceeded at a few research facilities such as the wind-tunnel laboratories of Stanford, the Massachusetts Institute of Technology, and the National Advisory Committee for Aeronautics, there was as yet no program of aerodynamics research supported by theoretical studies of the depth and sophistication that characteritzed Prandtl's and Kármán's works. In this chapter I trace American perceptions of German aeronautical research from the inception of theoretical aerodynamics around 1910 until Theodore von Kármán's first visits to Pasadena in 1926, which marked the beginning of a serious attempt to correct American weakness in the science and so, they argued, in the technology of flight.

In 1909, the *Scientific American Supplement* published an English-language extract of Felix Klein's recent account to the Aeronautical Association of Lower Saxony of research at Göttingen University.[2] He presented the report, a history as well as a blueprint for the future of the small wind tunnel at Göttingen, he said, to clear the record of "errors and misstatements" about the laboratory. But Klein's essay was as much a bid for wider recognition (and support) among aeronautical scientists, even among the Americans. The infant aerodynamics institute would soon receive the recognition Klein was seeking and, by 1922, accolades for the superior quality of its research. English and especially American reviews of the first two volumes of the hardbound report *Results of the Aerodynamics Research Institute* at Göttingen, published in 1921 and 1924, leave no doubt that many Americans admired German organization and productivity in research. These were the qualities to be fostered in American aerodynamics laboratories; the French and the German laboratories particularly were to be models of organization for the Americans, as the very first American references to European aeronautical science indicate.

As early as 1912, in a series of articles in three issues of the brand new *Aero Club of America Bulletin*, Albert Francis Zahm, an American pioneer of research in aerodynamics, argued the need for an aeronautics laboratory in this country.[3] Zahm complained that in the United States, as compared to "some localities, as in France . . . , the progress [in aeronautics] is halting, haphazard, and fortuitous,"[4] a lament that he and others would utter repeatedly in the next fifteen years. A later article of 1912 revealed that Zahm was particularly impressed with Gustave Eiffel's aerodynamics laboratory at Auteuil, near Paris, if mostly because Eiffel had invested his personal fortune in the installation: "He has set a noble example for emulation by some magnanimous American who would found a monument worthy of the highest praise and of great permanent value to his country."[5] Zahm was looking for a bequest to American aeronautics. In these early years of American aviation, when its promoters were seeking to broaden the basis of support, they saw advantage in pursuing aerodynamics studies at research institutes similar to those across the Atlantic, to guarantee similar excellence. Zahm was concerned to gain the consequent prestige of an academic association, and he begged American pursuit of an institute-model in these words:

> Wherever an appropriation for an aeronautical establishment may be placed, it is of cardinal importance that the directorate and personnel, as well as the endowment, shall be of the same high character as those found in the leading European countries, if not directly modeled after them, and it would doubtless enhance the prestige and efficiency of the institute to have it connected with an established institution having a reputation in the prosecution of theoretical and applied science.[6]

Having noted the weakness in American aeronautical science, as compared to the strength of the French, British, and German, Zahm, who had conducted his own wind-tunnel experiments since 1901, undertook a tour of Europe for the Navy and the Smithsonian Institution with J. C. Hunsaker, then an "Assistant Naval Constructor" who had been detailed to organize aeronautical training at the Massachusetts Institute of Technology. They visited the principal aeronautical laboratories of Europe, near London, Paris, and in Göttingen in August and September of 1913. Zahm published his official report for the Navy in the *Smithsonian Miscellaneous Collections* in 1914 (figure 14)[7] while Hunsaker published a popular account in the same year in *Flying* magazine.[8] Two years later Hunsaker's formal "Reports on Wind

Tunnel Experiments in Aerodynamics," based on the same tour, summarized the European research through 1915, and it too appeared in the *Smithsonian Miscellaneous Collections*.[9] Both Zahm's and Hunsaker's reports refrained from praising the European laboratories, but their very focus on Europe suggested the importance of these labs in establishing the direction of American research in aerodynamics. Zahm noted that Ludwig Prandtl's laboratory had begun on a small scale because of uncertainty of the practical value of wind-tunnel experiments, "though Prof. Prandtl has written some valuable theoretical investigations, and is reported to be undertaking large-scale experiments in the open air by use of a car on a level track, as at St. Cyr."[10] Hunsaker's was a report on the experiments and results of various laboratories, and mostly excluded judgments of their value.

Hunsaker's popular essay in *Flying* stood apart from the two official reports. It was clearly meant to garner support for the scientific study of aeronautics. The editor introduced the article with these words of advocacy: "From reading Mr. Hunsaker's notes it is apparent that in those countries where the greatest progress in aeronautics has been made a keen interest has been taken in the scientific side of the art. It is to be hoped, therefore, that our national laboratory and our greatest technical school may supply the research data and the engineers to apply it and that the United States may take a leading part in the future progress in aeronautics." Hunsaker himself noted the work in France, England, and Germany, and emphasized the advantages of the organization of the institute at Göttingen. "The staff of the laboratory has been composed principally of university students who were candidates for the doctor's degree. In this way much higher grade men have been had than could have been secured by mere employment. The character of the work done at Göttingen may be largely due to this extremely able personnel." The system of German academic study he thought enhanced the quality of research by providing cheap, skilled labor, a virtue that Guggenheim also would emphasize in 1926. And Robert Millikan too argued for this virtue and cultivated it at Caltech.[11] Hunsaker pressed for a similar scientific study of aerodynamics in America while underscoring the adolescence of American aeronautics with a rhetorical evocation of the father of American science: "It is reported that the worthy Franklin, on being asked what might be the use of the Montgolfier balloon, replied 'Of what use is a new-born babe?' Such progress in the conquest of the air has been made since Franklin's day that following his metaphor we may today inquire 'Of what use is a half-grown boy?' "[12]

Scientific, especially experimental, study of aeronautics in America gained sustenance from the founding in 1915 of the National Advisory Committee for Aeronautics, even if, in Richard Hallion's broad phrase, "until the passage of the Air Commerce Act of 1926, NACA's work was largely advisory."[13] But theoretical aerodynamics was almost totally lacking in America. This state of affairs is explained in part by the general (if not universal) aversion of American engineering schools to theoretical scientific studies at the expense of the practical. Not many theorists of the first rank were working before the First World War in American civil and mechanical engineering either. Beyond this, aeronautics was not even a recognized branch of engineering in the first quarter of this century. The lack of theory can be better understood by viewing Germany, not America, as the anomaly. There flight was treated earliest as a proper subject for engineering investigation at its "highest" level, that of theory, of the "applied science" characteristic of Göttingen. The English and French efforts in this regard were fewer, smaller, and, I believe, either more purely empirical or more distant from airplanes when mathematical. Finally, in America World War I retarded temporarily the growth of aerodynamic research. (In the long run, nevertheless, as with Prandtl's institute, it stimulated both research and interest in flying.)

After the war, despite antipathy to other things German, signs of American interest in German aerodynamic research appeared. The NACA resolved in November 1920 to hire as technical assistant one of Prandtl's students, the theorist Max Munk, who was at the time working for the Zeppelin Company in Bavaria.[14] Munk stayed with NACA for six years, contributing theories of flow around airships, of lift and induced drag of wings, and of moments and positions of centers of pressure on aerodynamic surfaces.[15] His responsibilities included supervising a growing program of experimental wind-tunnel research, which began with the opening of NACA's Langley Memorial Aeronautical Laboratory in 1920.

A spate of reports on the Göttingen institute appeared in English-language aeronautical journals after the First World War. Before the war British journals had appealed to both American and British aeronautical enthusiasts. Aware of developments on the continent, they served as transmitters of news of centers such as Göttingen to the New World. The British semitechnical journal *Engineering*, which had shown interest in German research by publishing as early as 1911 translated extracts from Prandtl's contributions to *Zeitschrift für Flugtechnik*, examined at length and with diagrams Prandtl's new aerodynamics laboratory, which had been completed during the war.[16] The warrant for this review was the appearance in 1921 of the first volume of Prandtl's *Results of the Aerodynamics Research Institute at Göttingen*, and *Engineering* summarized the introductory sections of that volume. Two other British journals, *Aeronautical Engineering*, the technical supplement to *The Aeroplane*, and *The Journal of the Royal Aeronautical Society*, a vehicle of advocacy but one held in high regard, undertook reviews based on the same work.[17] The latter, appearing in 1924, emphasized the "able direction" of Prandtl and noted that Göttingen was "the chief centre of aeronautical research in Germany." These reports were widely read by aviation enthusiasts in America as well.

Americans too reviewed the Göttingen laboratory. The popular magazine *Aviation* published a series of photographs of the institute and its wind tunnels in 1922 (figure 15).[18] Two years later, on the occasion of the appearance of Prandtl's second report, aerodynamics authority Alexander Klemin discussed the work in the same journal. Klemin, himself British educated but at NYU attuned to the problems of the working engineer, noted that the two volumes "contain a wealth of aerodynamic information, presented with great clarity and quite evidently based on the very soundest experimentation. While the work at Göttingen is scientific and based on modern aerodynamic theory, the experiments undertaken and the positive results obtained are of immediate practical value in design." Prandtl's work would be of use to the American engineers, he was suggesting. Klemin remarked also that Prandtl had incurred some delay in publishing because of the recent depreciation of the German mark, but happily the Göttingen institute had "got over its difficulties."[19] Aachen had not been so resilient.

The significance to American aeronautics of Prandtl's experimental work is further indicated by the contents of three NACA reports of 1920, 1921, and 1924.[20] They constitute a serial compilation of aerodynamic characteristics of airfoils used to design airplanes. Of 503 airfoil sections presented, 166 had been investigated at Göttingen, a number that far exceeded that of any nation other than Germany. That NACA reproduced the Göttingen airfoil data largely because they were readily available in print only underscores the visibility that Göttingen enjoyed in America. Indeed NACA also established a formal relationship with Prandtl in 1920, in which he was paid $800 for a report, "Applications of Modern Hydrodynamics to Aeronautics," which summarized the theories developed at Göttingen.[21]

Interest in the goals and methods of the Göttingen laboratory grew in the 1920s, especially because of the reviews of one respected American physicist and leader of the aeronautical community. Joseph S. Ames, professor of physics at The Johns Hopkins University and chairman of the Executive Committee of NACA, a man given to delivering speeches advocating scientific research on flight, presented one such polemic to the meeting of the section of physics and chemistry of the Franklin Institute on October 6, 1921. It was published subsequently with figures and photographs in the *Journal of the Franklin Institute* and reprinted in the *Smithsonian Report for 1922*,[22] and so it carried notable influence. Ames argued that navigation of the air consists of both an art of flying and a science of flight, and he wanted to emphasize the importance of the latter. The nature of aeronautics is essentially scientific, he asserted, and all progress in the art of flight must be based on its future scientific investigation; "the same methods must be applied [in aeronautics] as in any department of physics."

Experiments must be performed; a theory is evolved; deductions are made from the theory and tested by experiment; the theory is modified and improved, etc. During all the process knowledge is being gained, and the facts being made known help the designer of aircraft to make improvements in speed, in carrying power, in safety, in stability.

One most important fact should be emphasized, and this is that without the series of scientific studies just outlined not only would flight itself have been impossible, but also all progress in the art would cease. Scientific investigation forms the most important feature of aviation, and it can be conducted only by trained students.[23]

(It is striking that Ames has described no typical method of research of the period in the greatest centers of physics. In this discipline, of which Ames called aeronautics a department, the research program that reigned most exciting in 1921 was prosecuted in a very different manner. It was the quantum theory of atoms and radiation, a major counterinstance of Ames's formula, for it did not merely evolve out of experimental results in a kind of Baconian synthesis but embraced conceptual leaps beyond rational induction. In it coexisted apparently inconsistent ideas like electromagnetic waves and isolated quanta of light, indeterminacy in atomic events and spectroscopic regularity. Theory was not divorced from experiment to be sure, but often neither were they in good agreement. Instead Ames was calling upon propaganda that had been used to describe much of physics as it was pursued in America of the time, especially in his emphasis on the connections of "science" and applications,[24] and of applied physics of Germany, undertaken at the polytechnic institutes of Aachen, Hanover, and Munich, for example, but rarely at the universitites. The major exception in Germany, we know, was Prandtl's Aerodynamics Institute at the University of Göttingen, where such rhetoric had convinced the industrialists, through the Göttingen Association, to fund aerodynamic research. And so it would be logical that Göttingen should draw Ames's attention, as it did shortly.)

In order to accentuate the need for a discipline of aeronautical science and consequent funding of new researches, Ames depreciated the role of the individual, practical person. He argued: "Aeronautics is in no sense a function of an engineer or constructor or aviator, it is a branch of pure science. Those countries have developed the best airships and airplanes which have devoted the most thought, time and money to the underlying scientific studies." His words echoed those of the editor of *Flying* seven years before. But Ames singled out Germany and Prandtl among the Europeans for special citation, his only reference to a particular European center: "Great progress has been made in recent years by Prandtl and other German physicists by showing how a flow of air around an airfoil could be produced in an ideal frictionless gas similar to that observed in air, by imagining vortices or whirls in the gas." He mentioned that Prandtl and his disciples had been able to calculate the influence of one biplane wing on the other, and concluded with the words "and they have proceeded much further and made aeronautics into a beautiful theoretical science."[25]

Ames sounded enthusiastically supportive of Prandtl's kind of research, and indeed he was, as a later talk at the Centenary Celebration of the Founding of the Franklin Institute on September 17, 1924, underlines. There he concluded:

As a consequence of this theoretical work of Prandtl, [Prandtl's assistant] Betz and Munk, the science of aeronautics has at its disposal certain formulas which enable one to calculate the principal properties of an airplane wing of given profile and dimensions and also of any combination of wings. Furthermore, these formulas make it possible to discuss the effect of modifications in the wings, as, for instance, of changing the ailerons. It is true the formulas are based upon simplifying assumptions; but the results obtained from them are of the same order of accuracy as those obtained from tests in wind tunnels.[26]

An examination of NACA correspondence reveals the reason for Ames's support of studies of the "frictionless gas," an idealized approach that Prandtl had sometimes shunned, as with studies of the boundary layer. It was Prandtl's student Munk, whose work Ames knew better than Prandtl's and who pursued a classical (inviscid) hydrodynamic theory of wings and drag, as much as Prandtl whom Ames was supporting. In part Ames supported Munk simply because Munk was now a member of the NACA laboratory. Writing to J. C. Hunsaker on April 30, 1924, Ames had judged, "I think Munk's method of approach is much better than that of Prandtl. . . . In my judgment this is the most striking thing done in recent years in aeronautics." To an objection implied in the context, he wrote, "I agree with you in thinking that Prandtl and his associates gave a new life to hydrodynamics, but I do think that Munk has now started off on a new line which is much more interesting. Professor Bairstow of course believes in solving the equations taking into account viscosity, but I believe that for many years to come Munk's method will give us most of the information that we need."[27] Ames's concluding opinion would be proved an exaggeration. The works of Prandtl, Kármán, and Munk nevertheless represented new heights of postwar theoretical study, and Ames's talks were meant to raise appreciation of this German expertise in American aerodynamics.

Ames's passing reference to Professor L. Bairstow, occupant of the chair of aeronautics at the Imperial College of Science and Technology in London, is interesting. On at least two occasions in 1923–24, Bairstow delivered remarks in which he advocated the treatment of viscous effects in aerodynamic problems by use of Prandtl's boundary-layer theory (thus suggesting that the theory had been in some disuse, at least in England). The talks had fanned the smoldering priority dispute between Prandtl and the Englishman Frederick Lanchester. In the heatedly nationalistic postwar atmosphere a disputant in one of Bairstow's audiences asserted with outrage that the English neglect of Lanchester's theories had allowed English aerodynamics to fall behind that of Germany. He did not dispute the theory; instead he took issue with giving Prandtl full credit.[28] It is clear that Ames was *not* concerned with this English/German priority dispute here, for he preferred the approach of one German's to another's. He no doubt felt compelled to comment on Bairstow, however, because the English professor's remarks and the dispute had gained such prominence in recent months that no comment would have been conspicuous by its absence.

A final remark is in order on the significance of Munk's work in establishing the profession—as opposed to the corpus of knowledge—of theoretical aerodynamics in America. Two facts, taken together, suggest why his impact in this former respect was ephemeral compared to Kármán's. First, working for a government laboratory and with supervisory responsibility for much of its experimental work, Munk could not attract disciples to his theoretical methods, nor did he train more than a few students in them. Without disciples to spread the word, his work inevitably remained insular in America through the 1920s, of note only to the few who already studied theoretical aerodynamics. This situation was in marked contrast to that of Kármán, whose personable manner and world reputation brought students to Aachen and Caltech and who attracted large audiences to his entertaining, lucid talks. Moreover, Munk's works were notably abstruse by American standards; as George Lewis, executive officer of NACA, put it, "of such a highly scientific character that they are not appreciated by the average aeronautical designer, and can be appreciated only by those who have had a very extensive training in mathematics and physics." Lewis wanted popular versions of the theories written and even solicited Ames to write them.[29] But without a base of teaching and of intellectual exchange such as Kármán (whose work was equally difficult to understand) would have at Caltech, these efforts were futile. Munk's departure from NACA in 1927, after a reputed disagreement with Lewis and rebellion of his subordinate managers,[30] surely deterred further advancement of his ideas.

Just what was the relation of theoretical study in America to design of airplanes early in this century has not been examined, nor will I but touch on it here. In the 1930s, it seems, theory began to play a prominent if passive role in explaining the aerodynamic and structural behavior of an aircraft in an airstream. Occasionally that role became active, extending to designing features of aircraft, as in the case of the DC-3 at Caltech.[31] Design certainly relied heavily between 1910 and 1930 upon scientific empirical data like those from Göttingen and Stanford on airfoil sections and propellers.[32] But the relative newness of Prandtl's, Kármán's, and others' theories prevented their use in that period. (It is not in general characteristic of practitioners of engineering design to quickly adopt new theories; this would be especially so in aeronautical engineering, despite the apparent irony of conservatism, where an undisclosed deviation of the theory from nature might cost many lives and popular support of the adolescent industry in a spectacular crash.)

Indeed, the interesting question is what really motivated demand for theoretical studies in America. As Ames points out, one of the great attractions of Prandtl's theory was that it could provide numbers as near to those measured in wind tunnels as the experimental uncertainty itself—but theory could provide them at far less expense of time and money. Then too, a theory gave the designer a sense of "understanding" what was going on, and so offered a tool whereby he could sharpen his physical intuition of what behavior an aircraft would exhibit. Another motivating factor lay in that the academics like Ames and Millikan sensed that a true science of flight would not exist without an adequate theory to both explain the existing facts and generate new ones. Practical builder Donald Douglas might use the theory to advantage, but lacking this sense he would not support its development. The academics instead pushed themselves to the source of knowledge, for there lay their rewards, only justifying their search to the outside world by pointing to its applications. The prestige of the theorist in his profession was rarely mentioned in America, but its attractions nestled in the backs of minds of men such as Millikan, Epstein, Ames, and even Guggenheim.

Prandtl's and his students' successes, his conformity with Ames's conception of proper scientific investigation of flight, and the more widespread approval of the Göttingen institute suggest that still other American aeronautical supporters looked to Prandtl and his institute for advice, data, and most important for a paragon of a "scientific" laboratory of aerodynamics. And this I have noted in the 1926 travels of Guggenheim and Cone. Millikan also turned his attention to Göttingen. Yet he selected instead as his agent of transfer of the German institute of applied science Theodore von Kármán, the product but rival of Göttingen, to direct his laboratory.

8

CHOOSING A EUROPEAN AERONAUTICAL CONSULTANT FOR CALTECH

Ames's idea of employing students and research fellows to produce the highest quality research was shared by the leadership of Caltech. Begun in the departments of physics and chemistry around 1920, cooperative research of students, fellows, and professors would be extended to aeronautics as well by 1928.[1] There remained to be found an "investigator of the highest type" to set the new laboratory on course,[2] and eventually to direct it. In another of his remarkable letters, Robert Millikan wrote Harry Guggenheim in early July 1926 to inform him of progress in talks with William Durand of Stanford, of the upcoming meeting of the board of trustees of the institute, and to present his plans for bringing a European aerodynamicist to Caltech.[3] The letter in its entirety reads as follows:

July 7, 1926

Mr. Harry F. Guggenheim
598 Madison Avenue
New York City.

Dear Mr. Guggenheim:

I have just had a long talk with Dr. Durand, and have discussed this whole western airplane situation with him at some length, and I think he now appreciates quite fully the urgency and, indeed, the inevitableness of the airplane development in Southern California, and the desirability of having it directed along sound lines, especially in view of the fact that this has already become an airplane production center of the first importance.

Our Board of Trustees meets tomorrow, and I shall present your proposal, as contained in the memorandum that you handed to me at your home, to them and I think that they will take action to go ahead exactly in accordance with the plan proposed in my communication of May 14th, save for the modifications suggested by you in the financial set-up. I will, however, communicate with you their precise action as soon as it is taken.

With respect to the suggestion which you made as I left your house that we try to get Prandtl over here for a short time, I have talked the matter over at length with Epstein and Bateman. Both of them think that in view of Prandtl's advanced age and his somewhat impractical personality he would be far less useful to us than v. Karman, head of the aerodynamical laboratories at Aachen, and unquestionably the foremost of the younger aeronautical engineers of Germany; or G. I. Taylor, who occupies a similar position in Great Britain. Bateman and I both know Taylor fairly well personally, and Epstein is well acquainted with v. Karman (who, by the way, is a Hungarian in nationality). We have between us reached the conclusion, partially because of v. Karman's nationality and because of his representing in a special way the aeronautical developments on the European continent, that it would be well to try first to see if we cannot get him. As far as we ourselves are concerned, if we are going right ahead with the construction of a new aerodynamical laboratory, and also with a new design of the Merrill plane, which, by the way, is just now behaving most satisfactorily and showing great promise, we should like very much to have v. Karman pass on all these designs at the earliest possible moment. Epstein assures me that he will be completely familiar with everything that exists in Europe, and that a conference with him here will serve quite the same purpose as a visitation of all the European laboratories by one of us.

Since the chance of getting v. Karman over here will be very much greater if we can proceed at once than if we delay until after the next school year gets into swing, Epstein wrote him yesterday, at my suggestion, to tell him a bit of the situation and ask him if we were in such a position that if the opportunity came for him to make a trip to this country he could do so. My suggestion is that we offer him say $4,000 and ask him to sail for this country at once and spend three months here, getting back to Aachen about November 1st. If he could find it possible to stay still longer into the fall, so much the better; but Dr. Epstein thinks that the chance of his being able to embark upon this type of program would be very good if we could let him come at once, but he is rather doubtful if he could be away from his duties in Germany during a large part of the school year.

If this arrangement is made we should like to have v. Karman come at once to Pasadena so as to advise us on the aerodynamical laboratory,—this, of course, on the assumption that we make arrangements to go ahead with the "Daniel Guggenheim Graduate School of Aeronautics" at the Institute, as planned. On his way back he could visit Stanford and New York University, Boston Tech, Langley Field, McCook Field, and so on.

So as to avoid delay, Dr. Epstein has mentioned tentatively a plan something of this sort in his letter, suggesting that we might be able to secure the sum of about $4,000 for his expenses and services. As soon as we find definitely whether such a plan is agreed to we can cable him the information to the effect either that it has been possible to make the arrangements as suggested in the letter and ask him to come at once, or that it has not been. If the former decision is made we shall have saved three or four weeks of time, and if we decide not to get him no damage will have been done.

If your Board is interested in financing such a trip, would you let me know as soon as possible after the receipt of this. If not, I may be able to find some one out here who would be interested in doing it on behalf of the Institute alone. In case your Board wishes to get behind the enterprise and prefers Prandtl or G. I. Taylor, or indeed any one else, let me know and we will simply fit into your plan but cable v. Karman to the effect that it has been impossible to make the arrangements tentatively suggested in Epstein's letter, and we can then cable or write to Taylor or Prandtl or the other chosen person, whoever he may be, and start the negotiations de novo.

I hope you will not think that I have acted precipitously in this matter. Our course has been dictated by the fact that the chance of getting anybody is very much better if we can utilize the European summer holiday, which begins about July 25th, than it would be otherwise, and by the further fact that if we are going to move at the Institute we wish to move rapidly.

Hoping that I may hear from you with respect to these matters in the course of the next ten or eleven days, I am

Sincerely yours,

P.S. I hope that you and Mrs. Guggenheim can let us know beforehand about the time of your visit to this part of the world, which I understand you plan for the coming fall.

Millikan had discussed Guggenheim's suggestion to bring Prandtl to America with Paul Epstein, his man who knew mathematical physics in Germany,[4] and with Harry Bateman, the one who knew the field in England. Epstein, no doubt mindful of Prandtl's reputation for poor diplomacy and naivete and of Kármán's greater mathematical gifts, which he highly valued, and a friend of Kármán's, had urged Millikan to hire Kármán instead. Geoffrey I. Taylor, the English aerodynamicist, seems clearly to have been a second choice. Prandtl's operations at Göttingen would have impressed Millikan, of course. With separately directed institutes awarding doctorates in applied physical sciences, yet with an interdisciplinary atmosphere that was characterized by close relations of mathematicians, theoretical and experimental physicists (even if they did not always live harmoniously), Göttingen's was the kind of institution that Millikan was trying to create in America. But there were disadvantages to bringing a man like Prandtl to Caltech to organize the new laboratory, only some of which he mentioned to Guggenheim. Not the least of these was his German nationality and possibly his nationalistic ideology. He would surely be unappealing to many of Caltech's industrial and military patrons and, in any event, he would be difficult to entice from Germany.[5]

Kármán enjoyed a reputation as an excellent organizer, for he had turned Aachen into a major center of aerodynamic research, despite its less generous endowment. He was imbued with the methods of German aeronautical research and he was, to his credit, a Hungarian. In addition, Kármán was sufficiently younger at age forty-five than was Prandtl at fifty-one for the difference to work to his advantage; indeed, from Millikan's letter one might guess that Prandtl was a full generation older. It was true that Millikan considered the youthful excellence of his institute to be one of its greatest assets, "which brings with it the certainty of future opportunity."[6] Such opportunity was, after all, the raison d'être of the new aeronautical laboratory, and so Kármán gained the advantage. Finally, Kármán had lectured throughout Europe, and not only represented "in a special way the aeronautical developments on the European continent" through his research and teaching, but was familiar with the aeronautical laboratories in Europe, as Epstein emphasized. Millikan would have found his urbanity and internationalist outlook attractive, for he had cultivated these traits in himself.

If Guggenheim did not object, Kármán would be asked to consult. There was no mention of the directorship of the new Guggenheim School, but there can be no doubt that Millikan had considered Kármán's qualifications further for this position. It would be offered, as it was, once Kármán had shown his skills to Caltech and America. With characteristic zeal, Millikan had instructed Epstein to make a tentative proposal to Kármán, only afterward writing Guggenheim, expecting that Guggenheim would "not think that I have acted precipitously in this matter." The "European summer holiday" was only an excuse to hurry; the plan should simply not languish, especially since funding would soon be forthcoming and the design of the new School of Aeronautics was progressing.

Epstein had written to Kármán in his elegant hand and in German two days before.[7] Kármán, theretofore unaware of the plans for him, discovered them in the first sentence of Epstein's letter. "Allow me to state bluntly at the very outset the intention of my writing: the purpose of this letter is to explore with you whether you would possibly be inclined to spend several months with us this autumn in Pasadena." Epstein explained why the invitation came so late and what would be expected if Kármán should accept the offer. Without reference to a source, he mentioned that Millikan had just found the means to build the modern aerodynamics "institute" he had been planning. "It is quite clear that the consulting aid of a first class expert in this field would be of great use to us in the organization of a new department [of aeronautics]." Epstein asked if the institute could count on Kármán's being available if $4,000 were put at his disposal. (The sum was more than many full professors earned at Caltech in a year.[8]) Millikan had requested that a portion of his credit for the new school be applied to the consulting fee, "and it is probable, almost certain, that it will be granted."

Epstein noted that Bateman, an "excellent theoretical force for the direction of the mathematical side of the department," had a few young people of talent in research, but none had experience in organizing an experimental facility. He said nothing, however, of Bateman's qualities in this regard, perhaps because Bateman would be of limited use as a pure theorist. (Kármán later recalled that Bateman was a shy fellow,[9] and so probably he shunned or was never seriously considered for the job.) Perhaps, Epstein continued, Kármán could deliver several lectures to professional audiences on recent advances, "test the grounds" upon which the calculations for a new airplane design would be made, and perform similar additional duties. Millikan, Epstein, and Kármán also might discuss whether American students should be sent to Europe or if "a well-trained foreign employee [should] be appointed for a long period." (Either way, European strength would come to America. Epstein, himself a European import of 1921, had easily assimilated the view that America's aeronautical science lagged behind Europe's.) Epstein expressed also his personal wish that Kármán would decide to come. Others from Europe who had visited Caltech, among them Max Born, Peter Debye, and Paul Ehrenfest, had felt at home. And because the cost of travel to and from Aachen would amount to only about $1,500, Kármán would likely be able to save a lot of money. After talking again to Millikan, Epstein added the request that Kármán arrive as early as August 15. For their part, Epstein and Millikan would cable Kármán as soon as they knew if the proposal was definite.

A week later Harry Guggenheim telegraphed Millikan the following response to his letter of July 7: "Heartily approve of plan to have Kármán come to America on behalf of Fund to advise aeronautical educational institutions in this country."[10] Guggenheim had thereby broadened the purpose of Kármán's visit to a national tour of schools, to promote aerodynamic research and teaching. He had welcomed the idea of employing Kármán instead of Prandtl—but to a larger purpose, "for the benefit of advancing aeronautical science and education in the United States." Writing his West Coast assistant for publicity Ivy L. Lee, Guggenheim revealed, "My idea is that you might make a very interesting story for the press regarding Professor v. Karman's tour in America. Dr. Millikan is primarily interested in Professor v. Karman's visit to the California Institute, but my idea is to make him available for the country at large."[11] A week later he wired Millikan to ask if Kármán could deliver a series of lectures throughout the country.[12]

Guggenheim also wrote Millikan the same day to obtain his agreement to the contents of a letter to be sent to institutions where Kármán might speak, and he took the occasion to note that "Dr. Prandtl of course stands alone in the aeronautical world, but I heartily agree with you that for a practical visit such as you have in mind and which will fit in very nicely with our plans, Professor v. Karman is the right man."[13] Guggenheim had deferred to Millikan's judgment in regard to the needs of Caltech and American aeronautics. So Prandtl's candidacy was laid to rest, and Kármán, if he agreed, would be appointed.

But now Kármán faced a dilemma. Although he replied favorably to Millikan's cabled offer of July 20, he mentioned "some difficulty about proposed time."[14] It seems that Kármán had acquired two obligations for the coming term. The first was to the Kawanishi Manufacturing Company, an airplane builder which was planning to establish a research laboratory in Kobe, Japan. For this visit Kármán had already negotiated a six-month leave of absence from Aachen.[15] He considered two ways of dealing with the problem, as a letter of late July 1926 from his sister Josephine to his brother Miklos indicates. "Todor will go either to America or Japan at the end of August—30–31. About three weeks ago Todor got an unexpected invitation to go to California for two months. Now they telegraphed to Japan if it is all right if he arrives in Kobe (Japan) in December or if it must be the end of September."[16] He preferred to come to the United States first, and, according to his autobiographical account, increased the cost to Japan of his consultation by a factor of two, in order to "slip out of the obligation nicely."[17] The ploy failed when Kawanishi agreed, but Kármán succeeded in postponing the arrival date in Japan to December. He renegotiated his leave in Germany, extending it to a full year.

Kármán's motives are easy to discern. He would go first to Japan, to meet his obligation if he had to, but, as his sister remarked in the same letter, "this would not be pleasant because America is more scientific and means more as regards the future." Kármán understood the significance of the Caltech offer. He was eager to establish ties with the young but already prestigious California Institute, for reasons of both the opportunity in the United States and his situation in Germany.

Meanwhile negotiations between the Guggenheim Fund and

Stanford and Caltech had progressed to the point where, with a little prodding of Stanford, the two large grants could soon be announced.[18] But it would take until August 13, five days after the announcement in Pasadena,[19] before Kármán could say with certainty that he would be sailing to New York in early September. That evening Kármán transmitted his overdue plans to Epstein and to Millikan.[20] "I should prefer to visit the aerodynamical centres in the Eastern parts of the States before going to California," he proposed to Millikan, for thus he would be better apprised of the state of the art in America before he arrived at Caltech. He was exuberant at the prospect of coming to America. To Epstein he declared, "A visit in California, particularly in constant exchange with you and Mr. Millikan, seems to me very exciting." Kármán remarked that he was looking forward to discussing theoretical physics with Epstein, and he mentioned incidentally the great excitement that Schrödinger's new theory of wave mechanics had stimulated in Europe.

Events did not proceed quite as he had suggested. Kármán arrived in New York aboard the S.S. *Mauretania* on Friday, September 24.[21] After an "extraordinarily pleasant" weekend on Long Island at the home of Daniel Guggenheim,[22] he traveled on, not stopping in the cities he had mentioned to Epstein, but passing directly to Caltech, where he arrived at the end of September. There Kármán took up the duties that Epstein had outlined. Most notably, he redesigned the planned wind-tunnel laboratory after the method of Göttingen's, which Kármán had also adopted at Aachen. By his account there was some disagreement over the design. But as consultant, Kármán played the role of outside expert with authority; and this fact was one reason to change the mind of Arthur Klein, his momentary disputant.[23] In late October Kármán undertook his projected tour, lecturing at the University of Michigan, New York University, MIT, and other centres.[24] At NYU he met the same Alexander Klemin who had benefitted from Guggenheim's first grant and who had reviewed Prandtl's research in *Aviation* two years before. Kármán stopped also in Dayton and there spoke with Orville Wright, who described something of his own wind-tunnel experiments of 1902 to the surprised aerodynamicist.[25] In Washington, D. C., Kármán delivered a series of lectures and participated in an educational conference of university presidents and department heads called by the Guggenheim Fund, on December 10 and 11. The participants included Kármán, Klemin, E. P. Lesley of Stanford, Charles H. Chatfield of MIT, Felix W. Pawlowski of the University of Michigan, J. H. Parkin of the University of Toronto, and C. Magnusson of the University of Washington.[26] All represented once or future candidates for Guggenheim grants.

Kármán's lectures, delivered throughout the country, elicited Harry Guggenheim's high praise. Guggenheim wrote Kármán on December 9, "In all parts of the country where you have visited and where you have advised with our educators in aeronautics there has been a most grateful appreciation of your visit and the lectures in Washington, as you must full well know, are being received with tremendous enthusiasm and interest."[27] Perhaps a less interested review is that which Millikan sent Guggenheim, for though Millikan had his own reason to portray Kármán in a favorable light, he was more capable of judging the substance of Kármán's work. "Although von Karman's English is a little difficult to understand, he is a master of the subject and wonderfully skillful in presenting lucidly a subject which really involves a lot of advanced mathematics."[28] Thereafter, Kármán visited New York with his sister, but after seeing her off to France, returned by train to San Francisco.[29] In late December he sailed for Kobe,[30] where he spent the winter and spring (figure 16), returning finally to Aachen in August 1927.[31]

Kármán's American tour was most important for the interest in aerodynamics that it elicited. The plaudits Kármán received could only add to already manifold reasons for Kármán's excitement over the growth of aerodynamics at Caltech and the central role that he no doubt would play in it. Those reasons originated not just in America but also in the situation in which Kármán had found himself for the past five years in Germany. They deserve a closer look, especially in the light of the continuing correspondence between Kármán and Prandtl in Germany.

9

WHY KÁRMÁN LOOKED AWAY FROM AACHEN

Kármán's quick acceptance of Millikan's offer and the pains he took to rearrange his schedule in the autumn of 1926 indicate the enthusiasm with which he greeted this chance to work in the United States. Part of the reason for his eagerness lay manifestly in the positive aspects of the opportunity: the generosity of the offer, the excellence of Caltech, and Kármán's perception that the institute should have a promising future in aeronautical research. Although Kármán had no firm expectation that the consulting job would lead to a permanent connection, he no doubt considered the possibility. The situation in Pasadena in the late 1920s could not help but impress Kármán, for in many ways Caltech was the Göttingen of the New World. In their intellectual elitism and recognized prestige, in their pursuits of interaction between pure and applied science and in tapping the intellectual wealth of the former, and in the sheer abundance of their material resources, the two institutions bore common traits that appealed to Kármán.

But the question remains of why Kármán, unlike Prandtl, was willing to look beyond his institute and the German research community of which he was then a part. We are fortunate that a lengthy correspondence has been preserved, which reveals several incidents that, taken in sum, explain why Kármán was not altogether satisfied with his situation in Germany.

The Kármán–Prandtl correspondence in the Caltech Archives, which begins in April 1920, illuminates an event that could well have given Kármán pause in contemplating his future in German physics. Although Kármán had become internationally known for his work on the vortex street in 1912, with the interruption of World War I and the building of research facilities at his aerodynamics institute at Aachen, he would publish few papers until 1921. He had not yet committed all of his intellectual energy to applied physics. A letter from Prandtl to Kármán of April 23, 1920,[1] suggests that Kármán was considering pursuing a career in pure physics, perhaps in conjunction with his aerodynamic research. After the incident of Debye's replacement, described therein, however, he laid these ambitions aside.

Prandtl was writing to Kármán, he said, because David Hilbert, then the leader of mathematics at Göttingen, had nominated Kármán to succeed Peter Debye in the chair of experimental and theoretical physics at Göttingen. (Debye was a consummate leader of experimental and theoretical physics research, who had with his students produced numerous important discoveries since 1911, among them a quantum theory of the specific heats of solids. This theory had eclipsed in its reception by physicists the simultaneous, rival work of Max Born and Kármán himself. Debye's theory, though less exact than Born and Kármán's, in its simplicity better suited the needs of experimentalists and was more readily comprehended by quantum theorists.[2] The point was of significance in Kármán's career, as the following suggests). Prandtl admitted that he was partly responsible for Kármán not reaching the final list of successors to Debye. When asked about Kármán's work he had responded that most of it before the war was on topics in mechanics, hydrodynamics, and solid mechanics— except for the work with Born, in which, he said, "I view the main conclusion to be the calculation of the characteristic modes of vibration of crystals of masspoints, which again would lie in the area of mechanics." Prandtl not only questioned Kármán's credentials as a pure physicist but wondered also if Kármán was capable of directing the large physical institute at Göttingen. Evidently he had related these doubts to Hilbert.

Kármán's subsequent leadership in America belied Prandtl's opinion of his ability to administer a large institute. But Prandtl might be excused for misjudging his student's potential in this regard, especially because the job was in pure, not applied, physics. Indeed, Prandtl's net recommendation, measured by the standards of the institution and the time, was probably correct. But Prandtl had missed the point of Kármán's quantum work, a fact that must surely have galled Kármán. It was that Born and Kármán had *quantized* those modes of oscillation and so had come to an accurate and important theory of specific heats of solids. The calculation had been difficult and rested on new assumptions in quantum theory as well as on the mechanics of lattice vibrations. More qualified critics judged it a major contribution to quantum theory,[3] and Prandtl had underestimated its quality. For these reasons Kármán was no longer being considered.

But now, Prandtl continued, he had heard from Kármán's colleague Karl Pohlhausen, who was on leave in Göttingen, that Kármán was working in "modern physics." Prandtl invited Kármán to write him, to explain what he was undertaking, if he wished the position at Göttingen, and what scientific plans he would bring to the large institute. Born had already received the call but might decline, because his living situation in Frankfurt was quite comfortable. If there were enough new information forthcoming from Kármán, the issue could be reopened with the ministry in Berlin, which had formally to make the appointment.

Kármán's immediate response to Prandtl's condescension is not preserved. Born accepted the call, and two months later Prandtl wrote a few words to console Kármán.[4] He noted that Pohlhausen had spoken again enthusiastically of Kármán's research plans; Prandtl exclaimed, "Indeed, you have so much new in the works there that you really cannot leave Aachen and aerodynamics at all!" It would have been small consolation for one who aspired to Debye's chair, but it indicates to us at least that Kármán pushed on without interruption in aerodynamics. Prandtl's correspondence with Kármán remained cordial, even warm on occasion, as when after Kármán's fortieth birthday in May 1921 Prandtl changed the form of address in his letters from "Dear Colleague" to "Dear Kármán" on account of his reaching the so-called *Schwabenalter.*[5]

Kármán's career was coming to a certain resolution. Even if he regretted losing the opportunity to come to Göttingen, he let his feelings pass. In correspondence we begin to see him settling into a program of theoretical work in aerodynamics that Prandtl and other aerodynamicists recognized to be of the greatest importance. For example, Prandtl wrote on June 14, 1921, "I am anxious to see the manuscript that you have promised to send me,"[6] a work that turned out to be the important "On Laminar and Turbulent Friction" (see Chapter 6). And two years later, Prandtl sent a postcard to Kármán in response to a mailing of the outline of Kármán's theory of turbulence that he would present at the International Congress of Applied Mechanics in Delft in 1924. "I received your historic letter from Vienna today. Carefully examined, your conception is of moving simplicity, and leads in any case to a method that will bring us the solution."[7] Prandtl voiced some reservations, but his tone was one of respect and congratulation for creating the framework of a most important theory of turbulence that both men had sought. Not only did Kármán know his own worth; here was an indication, and not the first, that the most important of his peers acknowledged his excellence.

A second and more significant incident had colored his relations with Göttingen, in late 1922, however, thwarting for the moment his chances of reaching this mecca of applied mechanics and the pinnacle of academic status in Germany. It seems that as early as August 1920 Prandtl knew he would be called to the chair of applied mechanics at the Munich polytechnic, to succeed his teacher and father-in-law, the retiring August Föppl. Perhaps as further consolation for denying Kármán the chair of Debye, Prandtl wrote in that month that, should he go to Munich, Kármán would without doubt succeed him both as director of the Institute for Applied Mechanics and in Prandtl's chair.[8] At the bottom of the letter Kármán scribbled the names of seven mathematicians and physicists at Göttingen: Runge, [Edmund] Landau, Hilbert, [Richard] Courant, Born—then a line—then Kármán, and [James] Frank [sic]. He placed a check after the name of Hilbert, and checks before and after Courant and Born. These were likely the group Kármán expected to be congenial colleagues and those from whom he felt he would profit most. The name of the sometimes querulous, bulldog Robert Pohl, an experimental physicist, was notably absent, though he, like the others, headed a research institute. Prandtl had asked if Kármán would be interested in the applied mechanics post. There is no carbon copy of a reply in Kármán's files, as there are of some other letters to Prandtl, but these scribbles suggest the answer was positive. The Göttingen post was the center of aerodynamic science in Germany, and though Kármán was improving the status of Aachen, it could not compare with Prandtl's laboratories nor with the eclectic excellence of Göttingen's mathematics.

No further reference to the proposed move appears in the Kármán—Prandtl correspondence; but we may infer what happened in the following two years from letters of late 1922 to Kármán from Richard Courant and from Max Born. Prandtl had evidently vacillated about leaving. Yet in 1922 he seriously negotiated with Munich, and possible successors had already been considered. Kármán was not among them after all, and Courant felt that his friends owed him an explanation.[9] If Prandtl went to Munich, he related, the faculty would be called upon to fill the vacancy as an *extraordinariat*, that is, as a junior professorship. They of course would consider Kármán to be the only candidate in Prandtl's league but would argue that "a call to you must be abandoned because the position . . . would be an effective downgrading from your Aachen situation." With this subterfuge the faculty wanted to alter the character of the position to that of pure physics, for there were no other reasonable candidates in applied mechanics. Erwin Madelung, a theoretical physicist and also a Göttingen product, "who is ready to accept," had instead been suggested. Courant explained how he believed this situation had come into being and why Kármán's appointment was resisted.

Prandtl's call to Munich had not materialized until the last semester, when it appeared again as a likelihood. It seems that the Kaiser Wilhelm Institute, which included the wind-tunnel laboratory, would be directed by Prandtl's assistant Albert Betz, and so the vacancy would occur only in the post of applied mechanics and in the directorship of the institute of applied mechanics.[10] Hilbert, Born, and Courant had considered Kármán's candidacy as the only proper one, even if they had some doubts that Kármán would exchange his professional circle at Aachen for a more narrowly defined position at Göttingen. There appeared to be no hurry in the matter, and Courant was satisfied that he might discuss these problems with Kármán in person in the near future. But the issue became acute shortly before summer vacation. Doubts had grown among those who did not know Kármán, said Courant, as to the advisability of supporting his candidacy, "because of the anti-Semitic composition of the natural science faculty." "One can certainly understand it," he wrote perhaps sarcastically. "Born, Landau, Courant, Franck, Bernstein, perhaps soon Coehn, sit on the high faculty, from which are now excluded the less objectionable philologists."[11] He did not say it, but Courant was suggesting that those Jewish professors whom he listed might too be excluded if they pushed Kármán's candidacy. When Madelung's name was raised as a compromise, Prandtl, who "heard the most voices as Dean of the Faculty," agreed to consider the nomination. Suddenly, in August, an official notice from the ministry in Berlin suggested that the natural science faculty hurry, else the position fall to the department of agricultural machinery, a reversion to the field from which Klein had founded applied mechanics at Göttingen before the turn of the century.[12] With the added hurry it was clear to Courant that unanimity could be obtained and the post saved only by choosing Madelung. The appropriate letter from the representatives of the faculty was sent to the ministry, stating that they requested the appointment of a (pure) physicist, and naming Madelung. "Otherwise the position would probably have been lost completely." Prandtl, Hilbert, Courant, Born, and Runge signed the letter. They requested also that the ministry make public why Kármán was not called, presumably, because the call would have been a demotion and thus an insult. For Courant himself, he said, it had been a difficult decision to renounce Kármán's candidacy, and he added that the same was true for Born and Hilbert. Courant suggested that Kármán might still receive a call, "but on this I would like to write nothing." He acknowledged that Kármán might think that his friends had "left him in the lurch," and so he invited Kármán to Göttingen to discuss the matter. "Above all you should not misunderstand the attitude of Born. Born knows very well what you have meant for him, and would mean here."

If this confession was surprising in how much it said about the academic politics, and in particular the important role that capitulation of some Jewish professors to anti-Semitic repression played in this Prussian university in 1922, then Born's letter of less than a month later[13] was astounding for these and other reasons. Not just an acquaintance of or collaborator with Kármán, Born was a former roommate and counted himself Kármán's friend. His mode of writing, like that in letters to his friend Albert Einstein, reflected a different personality. It was self-analytical, indeed self-deprecating, and necessarily candid.

Acknowledging that Kármán should indisputably succeed Prandtl, Born admitted that he had refused to fight on Kármán's behalf. The problem was resistance that the larger philosophical faculty (of which the science group was a part) had mounted against appointing Born and Franck, both Jews, to professorships two years earlier. With five Jews in natural sciences, who were by no means united in their social and professional relations, he pointed out, their position in the faculty was "not simple"— indeed it would be further complicated by the nomination of another to a senior post.

Born sketched the positions of each of Landau, Franck, Courant, and Hilbert with respect to Kármán's nomination, suggesting that Born himself would have had to carry most of the burden of defending Kármán's appointment against the anti-Semites. He was clearly troubled for having stayed out of the fight, but equally convinced that he lacked the physical and emotional stamina to proceed singlemindedly. Born, the future Nobel laureate, acknowledged that it was Kármán who had brought him "into the proper physics when I threatened to be swallowed in the formalism of the theory of relativity. From you I learned also how one must work—despite your carelessness unworthy of imitation." But the question of the call was muddy. Franck had received two attractive offers, yet would remain in Göttingen if Madelung were called to the physical institute. And so Born signed the letter nominating Madelung to the Ministry of Education. (The complexity of Born's problem is suggested in a letter to Albert Einstein of August 6, 1922, in which Born remarked that "we have a lot of worries about professional appointments."[14] Prandtl's succession was but one place in a game of musical chairs that involved the possibilities of Pohl and Franck leaving and Madelung and Kármán coming.) Born sought to lighten Kármán's loss, suggesting that the time of special technical institutes at universities was largely past, despite the work of Felix Klein. But to Kármán the claim would soon ring false, for Prandtl's new hydrodynamics institute would open in just two and one-half years.

Born asked Kármán to write him what he thought of these events. The letter in sum was both a tribute to Kármán's abilities and a measure of the futility of resistance to German academic anti-Semitic prejudice that Born felt. It and Courant's letter might have left Kármán feeling less disappointed in the outward behavior of his colleagues than depressed at their stoic resignation to the oppression of *völkisch* politics. Or perhaps Kármán too could see no way out.

Kármán's response to Courant was silence, and his reply to Born, if there was one, is not available. It was not Kármán's style to write to colleagues in a soul-searching fashion, and so one must interpret from other sources how he felt about these developments at Göttingen. It is interesting that Kármán mentioned the contents of the letter from Born in a different context in his autobiography. Kármán remembered instead an oral comment of Born's after their joint study on the crystal lattice. " 'I enjoyed very much working with you, Kármán,' he said, 'and as a physicist I profited by following your way of solving theoretical problems. But fortunately for me I did not adopt your carelessness in handling data.' "[15] Although such a conversation may have taken place, the recollection is practically identical to and thus is clearly confused with Born's written acknowledgment in 1922 of the important role Kármán had played in his professional life. Kármán wrote in 1961 that he was genuinely hurt by the remark about his "carelessness." But, I submit, Kármán's dismay of decades later derived less from Born's observation that Kármán's methods were sloppy than from the hurt that the rest of the letter, now repressed, had brought him. Kármán could more easily fix on a claim of slovenly methods that was so obviously false and recall it as a source of the still-unhealed wound than reconstruct the circumstances of that time. One might even expunge the pain, as Kármán did, by drawing forth a moral, namely, that *Born* had been responsible for the mathematical details of the papers on specific heats and so had projected his errors onto Kármán.[16] This was the sin remembered, for it was a more forgivable one of fallen professional standards; the failure of heart was forgotten. But in 1926 Kármán remembered the affair too well.

Courant chided Kármán for his silence, in a letter of late January 1923.[17] The issue of Prandtl's succession was still alive and, Courant reported, "basically, the question of your call to Göttingen perhaps soon again must be broached here." Hilbert, Born, and he were working on Kármán's behalf, and so he begged Kármán to write when Kármán could come to Göttingen or another city to discuss the prospect, as was the custom. No further exchange ensued, however, for the issue was settled when Prandtl decided to remain in Göttingen to direct the new hydrodynamics institute that he would open in 1925.

Prandtl's role in the affair is unclear, although it seems to have been peripheral, in part because as the dean of the faculty he was in a position, if he so chose, of formal neutrality. Evidently he did.[18] His own future was still unclear and planning it surely provided such distraction as to remove him from the first line of dispute. Though Prandtl's name had appeared on a preliminary list of three candidates for a new chair of "Technical Mechanics" at the Munich polytechnic as early as July 28, 1920, his candidacy had waxed and waned twice through November of 1923. Föppl's chair would be divided into two posts of equal status, as the senate of the polytechnic argued to the Bavarian Ministry for Education and Culture, on account of "the great and manifold teaching responsibilities that comprise the field of technical mechanics." Thus both Dieter Thoma from a private firm in the Thuringian town of Gotha and Ludwig Prandtl, first and second on the list, would be called. Thoma was appointed to, and accepted, the new second post in August; but Prandtl, never an easy personality, would go only if *two* further appointments were made in mechanics. That, according to the registrarial records of the time, could not happen right away. So instead August Föppl's son Ludwig, the third candidate on the list, received the appointment in October 1921.[19]

As Courant had indicated, the appointment of Prandtl appeared, though theoretically in consideration, indefinitely post-

poned. The Munich records fail to show why, but in July of 1922 the faculty reconsidered his appointment, unusually without drawing up another list of three candidates. Asked if he was available, Prandtl declared himself ready and willing to take up direction of a laboratory of mechanics and to work with the younger Föppl. After protracted negotiations, Prandtl accepted the ministry's offer on December 30, 1922—that is, after Courant's and Born's long letters to Kármán. By then the worst of the affair with Kármán was over.[20]

But events turned quickly around in another six months. For Prandtl wrote the Bavarian ministry in June 1923 that the *Prussian* ministry together with the Kaiser Wilhelm Society for the Advancement of Sciences had awarded a generous grant to erect his new hydrodynamics institute at Göttingen—and he would have to consider it not just a counteroffer but a new call. By this semantic ruse Prandtl conjured an excuse to withdraw his previous acceptance of the Munich post and to remain in Göttingen. Munich clearly wanted Prandtl, however, and, receiving signals from Prandtl through the elder Föppl that the Berlin awards were not yet firm, sweetened the pot by adding a house for the professor as part of the proposed Bavarian appointment. Prandtl would have accepted had Berlin failed to meet its promises, which it almost did. But finally, with the relative stabilization of the economy by late 1923 and with further negotiation, Prandtl convinced himself that Berlin would provide the capital grant. He declined Munich conclusively on December 3.[21]

His decision came, by the way, somewhat to the annoyance of officials in Munich, as it surely did also to his father-in-law. In the Acts of the Bavarian Ministry of the Interior, a long "position paper" of September 26, 1924, outlined where technical subjects related to flight had been taught in greater Germany and how Bavaria, up till now backward in this regard, should advance aeronautical studies. On the copy of the paper preserved at the Bavarian State Archive in Munich, above the name of "Göt-

tingen" in a list of Prussian universities offering lectures in aeronautics, appears a penciled "x" and nearby the peevish marginal note of a reader: "Prandtl-Göttingen indeed had a professorship in the Munich Polytechnic, but he didn't come!"[22] Munich had so far failed to found a course or fill a chair specializing in technical problems of flight, the ministers thought, because Prandtl had strung them along for a year and a half.

Whatever external circumstances dictated Prandtl's footdragging, these did not include a concern for Kármán's position. They related instead to Berlin's vacillating willingness to support the new institute of hydrodynamics. And yet, in fact, the affair of Kármán's non-call to Göttingen did not destroy the personal relationship between Prandtl and Kármán, for there were many letters exchanged between the two in 1925 and 1926, some as cordial as they had been in previous years.[23] Kármán wrote a glowing tribute to Prandtl in 1925, on the occasion of Prandtl's fiftieth birthday.

But these small betrayals, if that is what they were, could only have convinced Kármán that he would be forever excluded from teaching at the greatest center of aerodynamics in Germany. Kármán would try to develop a new center with that status outside Göttingen.

By the mid-1920s Kármán was striving to make Aachen the center of attention for aerodynamics in Germany, even if at the same time external events worked against him. The evidence suggests that he had a legitimate claim on an expansion of the research facilities at Aachen and had the confidence of the Ministry of Education in Berlin. But the deteriorating financial situation of Germany in early 1923 militated against Kármán's enlarging his base of research at Aachen.

In late December 1923 Kármán lamented to Prandtl that in the economy "since Easter everything is possible"[24]; that the currently favorable monetary system of the *Rentenmark*, based on tangible assets, might again see a change for the worse. Rampant inflation, which had been checked only that autumn in a major economic reform surrounded by political uncertainty,[25] had made it extremely difficult in the last year for German scientists to travel and to acquire foreign publications. Financial austerity at Aachen was considerably worse than at Göttingen. The amount of research at Aachen was earlier to have increased, in conjunction with long-sought reform of Prussian polytechnic institutes; but the scarcity of Prussian educational funds had become especially acute in November 1922 because of the growth of inflation, and the ministry withheld all dispensable expenditures. The generally accurate commemorative history of the Aachen Polytechnic Institute notes two professors in particular, Kármán and Jaeger, whose proposals for using these educational funds were denied in the academic year 1922–23.[26] Perhaps as telling of the impact on Aachen were declines in the numbers of students in mathematical and physical disciplines in the years from 1921 through 1931. Enrollment had climbed slowly to ninety-four students in those fields in the year 1921–22, but soon the figures plummeted, from eighty-seven the next year to fifty-four the next to twenty-nine in 1924–25. The drops occurred in part because fewer could afford to study rather than work, in part also because of the passing of a demographic peak of veterans who had returned to school after the war. The numbers grew slowly again afterwards, but only in 1931 did they equal the preinflationary enrollment. Total enrollment at Aachen in these years never much exceeded 1000, less than one-half that at the Polytechnic at Hanover and less than one-quarter that at the Polytechnic in Berlin,[27] two more prestigious institutions which fared better in difficult circumstances.

Of course, universities showed similar declines in enrollment. The central University of Berlin, for example, comprised almost 12,000 students in 1921–22, but that number declined to 10,000 in 1923–24 and to 7300 in 1924–25. The Georg August University of Göttingen, however—an elite institution enjoying much private support and enrolling a larger proportion of students from wealthy backgrounds—felt less the effects of inflation and of demographic changes. With approximately 3100 students in 1921–22, the enrollment declined, but reached only to 2300 in the low year of 1924–25.[28] Göttingen had been a wealthy place independently even of Klein's efforts to bring it industrial support; for example, in 1910 its physics laboratory had ranked third in funding from government sources, behind only Berlin and Leipzig.[29] With a supreme reputation in the physical sciences and mathematics and a fine one in the humanistic disciplines, Göttingen would be among the last to suffer in a financial crisis.

Other statistics reveal more about the relative status and stability of the university at Göttingen and the polytechnic at Aachen. Total faculty at Göttingen in 1926 numbered 215, of whom ninety-eight were full professors. Göttingen's mathematical and natural science faculty, the largest at the university, comprised thirty-seven full professors.[30] This compared in the same year with Aachen's eight full professors in the so-called general science faculty, of whom the professor of aerodynamics was one.[31] Although the senior faculty at Aachen would be the last to feel the financial pinch, a number so small left little padding. And as enrollment diminished, so did Kármán's influence through teaching, just at the time when Prandtl's was growing with the construction of his new hydrodynamics institute. Kármán recalled in his autobiography that Prandtl had the advantage of an extensive experimental program at Göttingen.[32] It looked in 1925 as if Kármán would be unable soon to match that advantage.

But Kármán was not yet defeated in his German ambitions. In early 1925, he had received a call to the Polytechnic Institute of Dresden in the eastern German state of Saxony, and although he probably never seriously entertained the notion of exchanging Aachen for Dresden, he used the offer to his benefit. Kármán visited the Saxon Ministry of Education in early March 1925 and there outlined a proposal for the construction of a new aerodynamics institute at Dresden.[33] It included considerable help: three scientific assistants, two technical aids, an office aid, a mechanic, two technicians, and three apprentices. Kármán wanted guaranteed fees for lecturing of about 20,000 marks after taxes; rights to scarce lodging in those hard-pressed times—a dwelling with at least five rooms, a kitchen, and a bathroom; and the assumption by the state of the cost of this housing, if it were extra. Although his demands were exceedingly high, it is evident that he had the ministry in Dresden interested, for they wrote him about one month later, requesting that he send an overdue written outline of his proposal, as he had promised.[34]

In the interim, Kármán had visited France[35] and could not properly respond, but he may have wished not to do so immediately anyway. His demands constituted one of those negotiations whose clear purpose was to acquire an offer with which to improve the terms of his current position. For in a memorandum of May 25, 1925, the Prussian Ministry of Scholarship, Art, and Education proposed to Kármán improvements at Aachen, in an apparently generous offer.[36] Even in these months of tight finances, Kármán would receive the highest salary that a professor could earn then in Prussia. Furthermore, the administrator invited Kármán to propose formally his request for physical expansion of the aerodynamics institute in combination with the construction of a new Electro-Physical Institute at Aachen. And he made one more promise. Although a permanent subsidy of Kármán's laboratory from the Ministry of Transportation would be impossible, the German Research Laboratory of Aeronautics at Adlershof in Berlin, supported by that ministry, was considering awarding a series of research assignments. The Aerodynamics Institute of Aachen might be funded to solve problems in this program up to a level of 50,000 marks. The administrator stated that he was seeking a firm commitment and that he believed he would be able in the next few years to gain further research funds for the institute by similar means.

Kármán performed several calculations in the margins of the

memorandum. In one—on the final Dresden offer—he added some figures to a total of 14,793; in another, whose largest addend was 13,914, the Aachen salary, he reached a total of 17,518. Clearly the fees at Aachen were better. Kármán wrote Dresden to decline on July 16,[37] reminding the Saxon ministry of earlier misgivings about leaving an established research center and of his many connections with industry in the west of Germany, and noting that he had been given the opportunity to expand his institute at Aachen and to conduct research in the direction he wished. The last reason was significant in suggesting that previously his research had been somewhat restricted; freedom in research would be an additional reason for him to want to come to Caltech.

On the first of July, Kármán and the Prussian Ministry of Education had reached an agreement in which the ministry would undertake financing expansion of the Aerodynamics Institute "as soon as possible, in the next fiscal year at the latest."[38] This was the firm offer that freed Kármán to decline the Dresden chair. And yet, as Kurt Düwell's history of Aachen relates, the Prussian ministry failed for eight years to honor its later commitments to Aachen and Kármán to provide funds for teaching a course in statics of aircraft. As promised, the Prussian provincial legislature did approve a large appropriation for the new institute of electrophysics and the expansion of Kármán's institute in 1925, but, the history continues, the situation thereafter was characterized by ever fewer funds being made available for the Aachen Polytechnic Institute.[39] Expansion would be accompanied by neither more students nor the means to undertake a much larger program of research.

In fact, Kármán would have to bear a heavier teaching load despite decreased enrollment, which would lead in 1929 to well-founded fears within the Ministry of Education that Kármán might soon leave Germany if a second professorship of aerodynamics were not established.[40] Beginning in March 1926, letters of Kármán to Prandtl reveal that he felt he was overworked in both consulting and in administering his institute. "Unfortunately I hardly get to scientific research at all. The only thing I have done [lately] has to do with calculations on cantilever wings," he wrote.[41]

Finally, there were two reasons external to the progress of his science why Kármán could have felt by 1926 that his future might be brighter outside of Germany. The German nationalism that Kármán was obliged to suffer in the German aerodynamics community was the first. In a letter to Kármán of October 1923, Prandtl had asserted that he and Richard von Mises, editor of the *Zeitschrift für angewandte Mathematik und Mechanic* and the influential professor of applied mathematics at Berlin, had independently concluded that their names could not be listed beside "those belonging to the French Republic" as members of the planning committee for the International Congress of Applied Mechanics, to be convened in Delft in 1924.[42] Prandtl continued, "I should also genuinely wish to come into personal contact with no Frenchmen at this time," because of the French occupation of the industrial Ruhr region. Kármán responded that barring the French would not be appropriate, especially because they had officially declined to come and any French who participated would therefore be doing so in an unofficial capacity.[43] Kármán addressed Prandtl's misgivings with tact, but the suggestion of his reply from our vantage is one of distaste for dealing with national politics in science, the more so because Prandtl's reply of mid-December was far from conciliatory.[44]

Before May 1926, Mises and Prandtl again raised national political issues in correspondence with Peter Debye, who was helping to organize the Second International Congress of Applied Mechanics. Mises had warned Debye that if Rome were proposed as the site for the next meeting (which was held finally in Stockholm), the Germans would both oppose it and, if the decision came to Rome, they would boycott the meeting entirely. Debye, now in Zurich, asked Kármán to use his influence "in the most diplomatic manner possible" to overcome these animosities among the Germans.[45] Kármán might find it interesting, he noted, that only in the letters from Germany had there appeared political complaints with such petty objections. Letters from participants from other lands were completely free of them. Kármán responded to Debye, "I am completely of your opinion and naturally will speak with Prandtl and Mises when the occasion arises, so that all politics may be excluded."[46] Kármán did not relish dealing with this aspect of Prandtl, even if he unhesitatingly accepted the obligation from Debye. It was, after all, the very spirit of scientific internationalism that these congresses sought to embody, as their title emphasized. Unfortunately for Kármán, nevertheless, Prandtl and Mises spoke for the aerodynamics establishment of Germany. Upon aerodynamics, an applied science known for its strong connections to the military, rested the future of Germany's imperial expansion in Europe, and so such attitudes as Prandtl's and Mises's would likely reside in other supporters of German aeronautical science. Kármán confided to Millikan in 1928 that even he had helped German aeronautics in one war, but he did not relish preparing for a second.[47] In this respect he stood practically alone among the Germans.

Thus Kármán's status as an outsider, a Hungarian Jew, and an avowed internationalist was surely reason for his certain unease in Weimar Germany.[48] But this problematic situation was rendered more acute by the prospect of many of his German colleagues in physical science and mathematics publicly bowing to, in some cases incorporating in their research method, a neo-Romantic philosophy of life whose antithesis was the mechanical science and the technology that he pursued. In the words of historian of science Paul Forman, who has articulated the view in his work on Weimar culture, the dominant intellectual tendency of Weimar Germany was toward "reveling in crises and characterized by antagonism toward analytical rationality generally and toward the exact sciences and their technical applications particularly. Implicitly or explicitly, the scientist was the whipping boy of the incessant exhortations to spiritual renewal, while the concept—or the mere word—'causality' symbolized all that was odious in the scientific enterprise."[49] Kármán, who exerted special effort to keep abreast of theoretical physics and to maintain close ties with theoretical physicists,[50] would have felt alienated by these exhortations more than most other of the applied scientists. Not only were they irrelevant to a foreigner who did *not* equate the defeat of Germany in World War I and its aftermath to the Decline of the West, but they were subversive to his scientific enterprise. And to make matters worse, one could count among the "capitulators" to this spirit of the times some of Kármán's closest colleagues, among them Mises, Courant, and Born.[51] Born's remark of late 1922 about the decline of technical institutes at Göttingen, taken as evidence of just such assimilation of Spenglerian doctrine, would have estranged Kármán, for his commitment to applications was complete after about 1921. If Kármán was not thoroughly alienated by this negative aspect of science in Germany, the warmth of his welcome in Pasadena and American encouragement of his scientific pursuits put it in stark relief. Rebuffed twice by Göttingen, faced with decreasing enrollments and increasing teaching loads, not fully supported by the Prussian Ministry of Education, and finding now increasing political upheaval, antiscientific and anti-Semitic agitation, Kármán had ample reason to welcome a lucrative offer from the United States. Still, his acceptance was far from the permanent move to Caltech that he would only undertake eight years later. Perhaps he had in mind in 1926 extracting with the offer further concessions from Germany.

10

A PERMANENT CONNECTION

If Kármán was not sure of his own intentions when he arrived in Pasadena in early October 1926, he probably still was not sure at the end of 1929. The success of Kármán's American tour and his clear abilities quickly convinced Millikan, on the other hand, that he should eventually direct the new laboratory at the California Institute of Technology. Millikan would have further opportunity to see how Kármán would lead aeronautics at Caltech. Even before Kármán left for the Far East at the end of 1926, tentative arrangements had been made between Aachen and Caltech for a periodic exchange of Professors Epstein and Kármán.[1] Every two years each would spend three months lecturing at the affiliated institution. By mid-February 1927 the arrangement had been approved by the Prussian Ministry of Education, which was well disposed to the institution in America, which already held recognition and prestige in Germany.[2] The agreement would be invoked once by Kármán and twice by Epstein.

Epstein was the first to exercise it, residing in Aachen for the fall term of 1927, shortly after Kármán's return from his world tour that followed the first Pasadena visit.[3] Kármán, for his part, would return to Pasadena in September 1928. Thereafter he continued on a second tour around the world, visiting, among other places, Tokyo in January and the Hebrew University of Jerusalem in April 1929.[4] He returned again to Aachen in May 1929.

Kármán followed the progress of building the new Caltech aeronautical laboratory in the years 1928–1931 in long letters from Clark Millikan, Robert Millikan's son and an early doctoral candidate in aeronautics at Caltech. Millikan reported, for example, on April 2, 1928, that Stanford's E. P. Lesley had just completed design of the propeller for the institute's new wind tunnel (figure 17).[5] The correspondence reads like a series of progress reports to the man whom young Millikan considered the boss in absentia. Kármán's responses were friendly and replete with suggestions especially for hiring Europeans into the new school.[6] (Young Millikan, too, was to be treated with care, for despite his youth, to Kármán he was from the beginning heir-apparent.)

Meanwhile the elder Millikan wrote Kármán as early as January 1927 to express his delight at the confirmation of their exchange agreement with Aachen. He wished Kármán to be present at the finishing of the laboratory and the installation of equipment, about January 1928, and not so late as the fall quarter of 1928, as they had planned.[7] Millikan wrote twice more to press for Kármán's early arrival. Knowing that Kármán was a man who enjoyed physical comfort, Millikan emphasized that January was the beginning of the finest season in Pasadena and the worst in Europe.[8] Nevertheless, Kármán would be away from his duties at Aachen for a full year on his first world tour, and he would be unable to visit again until September 1928.[9]

Kármán remembered in his autobiography that one day during this second visit to Caltech in 1928 Millikan took him aside and "raised the question of the ultimate direction of the laboratory."[10] The day was December 6, 1928, for Millikan wrote Harry Guggenheim at once that he had talked to Kármán. Beginning his campaign for Guggenheim's blessing to hire him permanently, Millikan praised Kármán's "rare combination of exceptional theoretical grasp with practical insight into and physical approach to the problems of aeronautics." He spoke of the endorsement of his trusted theoretician Paul Epstein: "Epstein thinks there is no man in this country who is in his class in his knowledge and grasp of the whole field, or who would be of more outstanding importance for our country-wide aeronautical development if he could be intimately and fully identified with it." Millikan wanted to Americanize Kármán, to identify his personality and methods with the United States and in particular with Caltech. Kármán agreed, said Millikan, that the future of aeronautics was in the United States, because of the large distances between great centers of population here—Millikan's perennial aeronautical theme. And Millikan reported that Kármán had stated it was "distasteful to him, after his four years of aeronautical experience in the last war, to feel that he was engaged mainly in preparing for another one." (Kármán would have few qualms about preparing the United States for such a war; his misgivings lay rather in contributing to the military force of Germany.) Caltech's chief had posed the question if Kármán would come permanently to the United States. Millikan was unsure that he could meet Kármán's conditions of position and salary. For example, Kármán's institute received independent of salaries 20,000 marks from the government for military research, and presumably he would expect similar support here. Millikan was not asking for money from Guggenheim, he emphasized, but he wished to know if Guggenheim agreed that Kármán should be pursued to direct the Guggenheim Laboratory at the Institute.[11] A day later Guggenheim urged Millikan via telegram to seek funding to hire Kármán.[12]

The offer, and a purported similar one from William F. Durand of Stanford, were in Kármán's recollection "appealing and flattering," but he remembered his immediate response as negative. Kármán had confided to Millikan that Pasadena would be the only place in America to which he might wish to move; his declining Stanford was relatively easy.[13] Caltech was a different matter; Kármán would take it seriously, but probably his talks with Millikan never reached the stage of an offer in 1928. Yet Millikan had evidently not lost hope at Kármán's rebuff, if that is what it was, because on February 17, 1929, he wrote to

Guggenheim that he wanted to "discuss the von Kármán situation," presumably the questions of funding and winning Kármán.[14] Although the meeting did not occur as planned, Millikan evidently came to some understanding with Guggenheim, because in a letter of about March 1, 1929, Millikan formally offered Kármán the directorship.[15] The offer was generous and diplomatically posed, for Millikan knew there were factors in Kármán's deliberation that made him still reluctant to leave Germany.

It was above all his mother who would have to be convinced before Kármán could come to California. Kármán, and time, would slowly change her mind, and his own part of the decision would be helped as the terms were improved. In a letter of the summer or early autumn of 1929, Kármán's sister Josephine ("Pipö") reported to her brother the misgivings that their mother felt over the prospect of Kármán's moving to America and added her own advice: "She takes the Pasadena invitation with great repugnance. She does not consider the ocean, the long trip, and Millikan's personality as a paradise of the future . . . Mama said flatly that you will not leave Europe . . . Todorka do not decide anything but negotiate with the ministry first and don't write anything about it to your colleagues until you do . . . I think it is impossible to go to America completely; for the time being try Aachen."[16] It appears that sister Pipö too was resisting the move. If Kármán were to leave Germany, his sister and mother would accompany him, just as they had lived with him recently in the town of Vaals, the Dutch suburb of Aachen.[17] Their intransigence and Kármán's lingering loyalties to Aachen were surely part of the cause of his vacillation. He remembered his rejection of Millikan's initial offer of December 1928 with the qualifier, "I must admit, though, that I didn't turn down the offers without serious reflection."[18] Evidently not, for he accepted the Caltech position in October 1929. But before he came around, Millikan and Guggenheim had to find further inducements.

Kármán had not responded to Millikan's first written offer by mid-June, when Guggenheim invited Millikan to confer with him at Guggenheim's family estate in Port Washington, New York. Besides the two, Emory S. Land, the successor of Cone as vice president of the Guggenheim Fund and J. C. Hunsaker, now of MIT, attended the June 14 meeting, which was called to determine how the fund might best support a laboratory for the study of lighter-than-air ships.[19] Several years of negotiation with the University of Akron and other institutions had aimed to found the laboratory, but only at this meeting did representatives of the fund concur in joining with the Akron institution and Caltech to establish a "Daniel Guggenheim Airship Institute" at the Akron airport. Kármán would be brought to this country, not only to head the School of Aeronautics at the Institute, but also to direct the Airship Institute from Pasadena. Millikan accepted a charge, though apparently with some reluctance, to "take leadership in Airship Research."[20] It was Hunsaker, as consultant the guiding force for airship research in the fund, who had insisted that Kármán direct it.[21] The conferees dispatched telegrams to Kármán and to Epstein, who was visiting Aachen at the time.[22] Kármán would be offered the directorship at Akron in addition to the Caltech position. In the former, he would be a nonresident director, providing "leadership" from his office in Pasadena. Epstein, again the go-between, telegraphed Millikan that Kármán would now likely take the jobs. As Millikan subsequently reported Epstein's message to Guggenheim, employment at Caltech would "assure him greater freedom from consulting demands than he has had in Europe,"[23] the reason that the Berlin ministry too recognized as a motive for Kármán's move to the United States.

On July 6, Millikan wrote Kármán a detailed letter of invitation in which he set out the terms of the appointments, adding that Kármán would be happier in Pasadena than in Aachen, as would his mother and sister. Kármán would receive $10,000 per year for life to direct the Guggenheim Aeronautical Laboratory in Pasadena and an additional annual $2,000 for the extra responsibility of and transportation to the Airship Institute.[24]

At the prevailing rate of exchange (about $.24 to the Reichsmark in 1929)[25] the American salary was three times greater than the sums Prussia had paid him since 1925. Even allowing that the Aachen income was tax free and that Kármán's cost of living in Germany (and Holland where he resided) stood well below what it would be in America, the Caltech financial rewards surpassed those at Aachen. Millikan had improved the offer, and Kármán was tempted; but he would not accept for the fall term of 1929. William Durand, who visited Kármán in July 1929, became like Millikan convinced that Kármán's reluctance derived from his family obligations, and he reported as much to Harry Guggenheim.[26] Guggenheim and Millikan each implored Kármán to accept.

Harry Guggenheim wrote a long and gracious letter in early August, formally offering the position of head of the Daniel Guggenheim Airship Institute but reassuring Kármán that it would absorb little of his time. Guggenheim listed the reasons Kármán ought to come to America: "upon your appearance in America you will immediately become the Dean of and have an unrivaled position in your profession in this country." As for the status of this country and of the institute in particular, "I feel that the California Institute of Technology, under Dr. Millikan's direction, is going to do very great things in a great section of a very great country." In Guggenheim's view the jobs would offer him greater freedom with time and money than in Germany: "Most important of all, I think would be the untrammeled opportunity that you would have to express yourself and bring to fruition these contributions that are constituting your life work." Guggenheim concluded with a gesture, asking if he might "have the pleasure of greeting you when you pass through New York."[27]

By his emphasis we may suppose that Guggenheim was most sensitive to Kármán's need for the prestige of the position. In his reference to freedom of inquiry Guggenheim was responding in line with Epstein's view of Kármán's heavy obligations in consulting, perhaps also in accordance with his understanding of the political environment of Germany in the late Weimar period. The greeting he offered from one aristocrat to another.

Millikan's letter, written late in August, more personally argued the advantages of the region around Pasadena and the expected comforts for his family. Southern California was the "logical place for your efforts and your influence during the next twenty years; and I am altogether sure that your mother and sister will be quite as much at home in it after you get here as will you." He reiterated the terms, emphasized the security, and expressed his hope that Kármán would make definite the "prospect held out in your cable of July 7th, that you will start your duties here next spring term." Would Kármán cable him to this effect as soon as possible?[28] Harry Guggenheim, meanwhile, had asked Hunsaker to request the German aerodynamicist Karl Arnstein of Goodyear–Zeppelin in Akron "to add his encouragement to get von Kármán to come to America."[29] Epstein too would stay in touch and several times added his voice to the chorus of persuasion.[30]

Kármán remained uncommitted until October. In the interim there evidently was silence between Kármán and the Americans, for no further letters appear in the Kármán, Millikan, or Guggenheim papers until October 9. An undated draft of a telegram from Kármán to Millikan suggests he put them off: "Could not move Pasadena September hope make it possible start duties next spring term. However, permit me postpone somewhat final acceptance. Detailed letter follows."[31] It is no doubt the telegram of July 7, to which Millikan had referred in his August letter. Only on October 9 did Kármán write the detailed letter he promised, in which he apologized for the delay, which he accounted for by citing (somewhat vaguely) continued deliberations with the Ministry and his mother's ill health.[32]

Millikan had been patient, even though in early September Guggenheim had given notice to the board that the fund would soon be dissolved.[33] But on October 11 he received word that the proposed agreement, long in preparation, between Caltech, Akron University, and the fund, was ready for signature. Harry Guggenheim wanted it consummated before he left the United States to assume his duties as new ambassador to Cuba,[34] and thus for Caltech's chief the arrangements took on new urgency. On October 18, Millikan sent Kármán a firm telegram setting out the offer including a "total annual budget under your control here [of] fifty thousand dollars, with prospects of even larger support" and requesting his "definite acceptance of directorship before writing Guggenheim" of Caltech's acceptance of the Akron agreement. Two days later Kármán responded that he had "decided in [his] own mind" to accept but he could not do so formally until November because of a promise to the Ministry.[35] It was the firmest statement Millikan would receive until Kármán arrived in April 1930.

In fact, there was more reason for Kármán's delay than a mere promise to the Prussian Ministry of Education, reason that he did not offer Millikan. Kármán was exploiting the Caltech offer in negotiations in early October to establish a second professorship of aeronautics at Aachen, in part to relieve him of his teaching burden. The ministry rejected the proposition, for state policy decreed near the end of that first depression year that no new chairs could be created in Prussian polytechnics. The Prussian ministry knew, however, that otherwise Kármán might soon leave, and it raised the prospect in mid-November of establishing a chair in Berlin—clearly a more prestigious position than at Aachen—with Kármán to occupy it.[36] But this too would not come to pass, and even though by mid-December his mind was still not completely settled, Kármán knew that he would be going to the United States. He wrote his friend Jan Burgers in Delft in December, "My fate still lies in the hands of the Gods; yet my mother is gradually getting used to the idea of Pasadena, so that in the end Pasadena will be the upshot at least for several years."[37] Three days later salaries of Prussian professors were reduced 15 percent because of the worldwide financial crash,[38] and the chance for an arrangement with Berlin was rendered very remote.

In response to Kármán's equivocal acceptance, Guggenheim had remarked to Millikan, "obviously our friend goes into the water inch by inch."[39] Even then he could not know how slowly Kármán was to settle in Pasadena. Kármán did come to America in April 1930 and was greeted warmly by his colleagues in Pasadena (figure 18). Epstein wrote him a short greeting on April 5: "It is a great pleasure for me to welcome you in America, and the prospect of having you lastingly as a colleague causes me great satisfaction."[40] But he would not yet take up the full-time position. Kármán immediately arranged to spend winters at Caltech and return to Aachen for the German summer semester of April through September.[41] In this way he would retain both his chair in Aachen and the directorships of the two facilities in America.

The arrangement in Germany, however, proved tenuous. After several weeks in Pasadena, which culminated in Kármán's appointment on May 31 to the half-time and half-salaried position in America, he traveled again to New York in June and then to Aachen in July.[42] Kármán returned to Pasadena in October 1930 (figure 19). He taught at Aachen in the summer semesters of 1931 and 1932,[43] but never thereafter.

A continuing correspondence with Prandtl in those years, carried on for the most part when both were in Germany, betrays no major strains between Kármán and the German colleagues in general or Prandtl in specific until late 1932. On April 24, 1931, for example, Prandtl wrote to Kármán, "You are awaited here (that is, in Germany!) with longing," and spoke of being pleased if Kármán could come to Göttingen to discuss a problem that had arisen in the German Research Council.[44] But even then there were lesser strains. As Kármán remembered, the Nazi party gained increasing influence at the Aachen polytechnic, an impression confirmed by Kurt Düwell's history of Aachen.[45] In August 1930, just after he had returned to Aachen, where he arranged for at least two of Prandtl's protégés to come to Caltech on temporary appointments, Kármán expressed to Clark Millikan a certain misgiving about "the large number of people that we are obtaining from Germany." He wanted to leave room for an American or a Briton, but none were as valuable as the Germans, he declared.[46] Two years later the worldwide economic depression had deepened, not least in Germany, and a certain petulent concern showed through the polite gloss of a letter from Prandtl to Kármán of September 29, 1932. Prandtl suggested Kármán keep his assistant Walther Tollmien as long as he could, for incomes (and means of supporting researchers) in Germany had severely shrunk.[47]

The National Socialists came to power on January 30, 1933. At first, their anti-Semitic dogma and the racial laws of early March had no impact on Jewish professors at Aachen. But on March 30 three Jews were suspended from giving examinations. Several more were forced to take leaves of absence on April 28.[48] Kármán, had he been in Aachen, might not have been among them, since his civil service dated to before the war, a qualification for exemption; but blanket suspensions of all Jews, regardless of the duration of their service, were usual when the laws were first enforced, and it is more likely that he too would have been an immediate victim.[49] He requested an extension of his usual six-month leave from Aachen on April 15, an extension that was granted through the summer semester of 1933. The Ministry of Education insisted, however, in an official memorandum to the rector of Aachen dated May 4, 1933, that Kármán take up his teaching in the winter semester of 1933–34.[50] Kármán remained in Pasadena that winter. With bureaucratic finality, the representative of the Ministry demanded on January 27, 1934, that Kármán either resume his service at Aachen immediately, or resign at the cost of his pension and other benefits.[51] Regretfully Kármán wrote his resignation, which included this disdainful declaration: "I hope that you will be able to do for German science in the next years as much as you have done in this year for foreign science."[52] Exercising the open offer from Caltech that allowed him to assume full-time duties there at his convenience, Kármán continued the year 1934 as a permanent resident of the United States.

In response to an inquiry from Prandtl about his future, Kármán had written icily in August 1933, "Thank you for your interest regarding my plans. As you mention, I got a short letter from Berlin suggesting that I take up my activities over there in the fall. I do not think I will do this; I find my situation here quite satisfactory. The German academic life has some advantages, for instance a definitely better beer than here, but I think you will agree with me that this is not sufficient reason for me to neglect the disadvantages."[53] The letter was the first he wrote to Prandtl in English. It was to show that from now on America would be his home.

In a story full of ironies, one of the greatest was that Germany drove Kármán away only partly intentionally. Unlike much pure physics—Einstein's theory of relativity, for example, which Nazi sympathizers Philipp Lenard and Johannes Stark and their followers vilified—aerodynamics was obviously needed to advance the aggressive plans of the Third Reich. Germany's new Air Minister may have been aware that the loss of Kármán would be unfortunate.[54] Kármán could have avoided the anti-Semitic actions in their early stages if high-level officials intervened, and there were examples of Jews who retained key positions. The very tenacity with which the Ministry of Education sought to induce Kármán's return suggests its motives were more than perverse.

Another irony lay in the very backwardness in research in theoretical aerodynamics in the United States (combined with Guggenheim's crucial willingness to remedy this situation with grants to improve aeronautical education), which impelled rapid advance of the American schools to world class in aeronautical teaching by 1940. Under these circumstances Kármán found the opportunity to eclipse his mentor in influence, which he did at least by the end of the Second World War.

One could argue that the role of Caltech made little difference to American aeronautical interests, for Kármán's position in Germany and American readiness to bring academic aerodynamics to America would have attracted him here anyway. But his familial relations and the character of his science suggest the contrary. He sought to satisfy the wishes of his family, and he sought a place where his own aspirations and needs would be fulfilled. These he found in Pasadena, much because of the favors, hospitality, and solicitations of Guggenheim, Millikan, and Epstein, the Caltech advocates. In Pasadena he received friendship, as well as respect and support for his ability, more than for his position.[55] No other American educational institution of the time would have offered him such unbridled freedom to pursue research in theoretical aerodynamics.

Historians have lately emphasized that American science was pregnant in the early 1920s; that, for example, if America could boast only a few superb physicists in 1920 (and no theorists meeting the German standards of excellence), by 1935 the abundant material resources and heightened interest among American intellectual leaders and the immigration of scores of scientists had caused a burst in research in this country and precipitated its world dominance in physical science and technology.[56] Kármán saw that potential in 1926; he sought to bind himself to the scientific future of the country then, and he took note of many of his colleagues—among them Paul Epstein, Karl Arnstein, Max Munk, and even Prandtl—visiting or moving to the United States. The wealth and the flexibility of American society provided Kármán with a midlife opportunity to reach the height of his profession, as he thought he deserved, and he took it.

Kármán closed his deal with Caltech less than two weeks before the American stock market crash. Because the Guggenheim Fund, whose awards ended in 1930, remained sound, as did Caltech's endowment, Kármán's position remained secure and his salary was undiminished. But he had negotiated a better arrangement than he might have only months later.[57] For many years he remained the third highest paid employee at Caltech. The reason for the high salary was in part purely economic: his work brought financial support from outside the institute, from the aviation industry and the military. This was not the case in sciences such as physics. Aerodynamics was an academic discipline that, because of its relation to aircraft, might also generate popular enthusiasm if presented properly. It was one of Kármán's skills that he could excite audiences and individuals through clear, witty lectures and talks.

The military support of Kármán's research bears special emphasis. When Kármán joined Prandtl's institute in 1908, Göttingen had just become the major research center of a new industrial, academic, and military alliance. Created by Felix Klein and his collaborators and backed by the kaiser, this alliance (and specifically Krupp, Klein, and Prandtl) had funded Kármán's earliest studies there, which were of use to the military.[58]

Still stronger ties were to come. In the early years of the First World War, as a Hungarian Army Lieutenant, Kármán had founded and directed a research laboratory for the Austro-Hungarian Air Force (figure 20). There he studied propellers and armament for airplanes and attracted considerable attention for his development of a tethered helicopter in 1917.[59] After the war,

Kármán had sat on the board of directors of the Fokker aircraft company, which also included prominent German military figures (figure 21). He had been a supporter of the student glider competitions, helping to design Aachen's winning entries and so to train the young pilots who would be Germany's bootleg air force in the twenties (figure 22). He had worked for the German Ministry of Transportation from 1929 through 1933. There he established a review committee of scientists to try to dispel Allied suspicion of German aeronautical research, which indeed would have a military purpose.

In the first phase of his association with Caltech, between 1926 and 1934, Kármán's military work was minimal. But ties were inevitable in the late 1930s, as the generals came to recognize him as a dependable booster of American aeronautics and a sympathetic and experienced consultant to the military. In 1936 he met General Henry H. Arnold of the army air corps, and by 1939 he was being called to Washington as consultant on how the air corps should carry out experimental research, in meetings that led to the building of a large, high-speed wind tunnel at Wright Field in Ohio. In the same year of 1936 he became consultant to the Army Bureau of Ordnance and a member of its advisory committee, consulting on aerodynamic characteristics of projectiles at supersonic speeds. In the war years he developed jet-assisted take-off (JATO) reaction motors, under contract to the engineering division at Wright Field (figure 23). The JATO development brought him an order from the navy for an adaptation of the invention to aircraft taking off from carriers. Kármán also produced for the army a preliminary design of a supersonic airplane in 1943. In what he must have counted the peak of his "military" career, Kármán attained the "simulated rank of major general" in order to interrogate German aerodynamicists in Spring 1945. This role he "enjoyed very much," for it carried both material and psychological benefits.[60]

The affection with which Kármán and the military embraced

was so extraordinarily different from the usual relations of scientists and generals at that time, if such existed at all, that it invites a search for special reasons. These were rooted firmly, I think, in Kármán's psyche. Certainly he appreciated government generosity in supporting his scientific research. But, more important, Kármán found the military soothingly predictable and satisfying of his need for recognition. If the German military establishment had treated him adequately after the first war, the U.S. Army embraced him as an exalted professor and enlisted him to perform no less a task than to plan the future of military aviation. Kármán, considering himself an outsider, observed and approved of Arnold's effective use of outside consultants to advise the air corps.[61] But one need not overemphasize Arnold's special ability to gratify Kármán's ego. There were many other friendly cooperations between generals and the aerodynamicist, which were accelerated by his direction of the Air Force Scientific Advisory Board in the late 1940s. Rather, Kármán found something in general flattering and attractive about the military's appreciation.

Kármán also found the military the easiest group to convince to support scientific research in his field. With them more than all others the arguments of scientific research as a foundation for technological development hit home, and they derived political advantage from an association with the acknowledged supreme aerodynamic expert of the nation. Defending the association was easy, in part because "scientific research" to the military as to Kármán encompassed a broad field of endeavors, from the obviously beneficial troubleshooting of an instability in the P-38 to the theoretical description of a fluid flowing past an obstacle at supersonic speeds. Kármán recognized that it was not usually necessary to make explicit connections between aircraft development and theoretical creation; for most military men, indeed most everyone, could not grasp those technical connections anyway. It was sufficient merely to provide the benefits to aircraft of doing basic research with due credit to their source. If connections existed between improving planes and solving problems of flow in the abstract, as Kármán genuinely believed they did, these he would record. But he and his disciples remained the keepers of the covenant, the wizards of aerodynamics, and the army treated them as such.

In his life story, Kármán suggested both these material and psychological reasons for his relationship with the military.[62] His success with this group, and his long-held view that as a group the military could be exploited to further the goal of scientific internationalism (with America, of course, at the top of the new international community), appealed to Guggenheim in the late 1920s. So did the material benefits of the association appeal to Millikan. Despite Millikan's antipathy to government support of science,[63] this relationship conformed to Millikan's larger concept of Caltech as an elite center of American science, supported by a broad economic base of industrial benefactors.

With Millikan's blessing, Kármán forged early links between industry and the military in America, which help bind together what is now called the military-industrial complex.

The degree to which Millikan succeeded in interesting persons outside of academe to support research at Caltech is comparable to that of Klein at Göttingen two decades earlier. Industrial and academic alliances of their sorts were unusual in both Germany and America and were more readily accomplished in technological fields like aeronautics than in the purely scientific. So Klein and Millikan would seek the best researchers in a field like Kármán's to build a bridge between engineering and academics while pursuing the interests of pure research. Klein and Millikan attracted support from industrialists with tacit promises of tangible, if long-term, benefits from the resulting knowledge, and, although some or much of their financial means were shunted to the "pure" sciences, it was the benefits of the applied that industry tallied. Caltech, like Göttingen, first succeeded in collecting industrial support and then in perpetuating the persuasion by taking care to conjure these noncharitable benefits to the industries. An applied science like Kármán's, therefore, was a valuable element of Caltech's total support; in its benefit to industry it added credence to Millikan's claims, and it carried the sophistication and values of a truly academic science, which were necessary to permanently seat aerodynamics at academically elitist Caltech. Thus Kármán's worth to Caltech better than equaled his salary. That worth was high also in relation to what other academic institutions that failed to seek industrial alliances might derive from Kármán's gifts.

As Harry Guggenheim and Robert Millikan had hoped, Kármán was virtually Americanized by 1934, as his letter to Prandtl of August 1933 suggested. Thereafter he represented the interests of America in international scientific and military organizations. He had brought his training in theoretical aerodynamics to the United States. Now, with his allegiance shifted to his new home, he gave aerodynamics an American brand and America international prestige in the field. This was Harry Guggenheim's ulterior motive for flattering Kármán about his 1926 series of lectures: he had wanted still more for America from the Hungarian scientist. From the first, Guggenheim had sought excellence in aeronautical teaching and research. He knew of an existing abundance of talent and material resources in the United States; but he and Millikan had looked to Kármán to exploit these resources, to organize them in the manner of the Germans, for the pursuit and transfer of the academic, mathematical science of theoretical aerodynamics and the engineering of airplanes (figure 24). This goal was a small part of Guggenheim's vision of American dominance of aviation, a dominance that itself would be but one means to world economic hegemony.

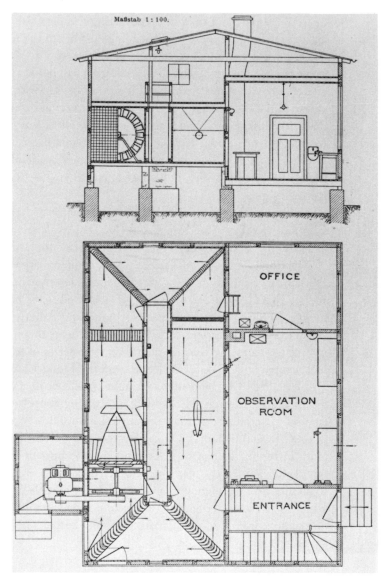

Maßstab 1 : 100.

OFFICE

OBSERVATION ROOM

ENTRANCE

14 Detailed diagram of the first
wind-tunnel laboratory at Göttingen.
From A. F. Zahm's 1914 review of
European aeronautical laboratories in
the *Smithsonian Miscellaneous Collections*.

15 Photo survey of the new
laboratory at Göttingen,
published in 1922 in *Aviation*
magazine.

16 Despite attempts to extricate himself from the commitment to consult
for a Japanese aircraft builder, Kármán seemed to thrive during his 1927
visit to Tokyo and Kobe.

17 Robert Millikan shows off the new
10-foot wind tunnel at Caltech, 1930.
Bettmann Archive.

18 Kármán in a dapper pose as new
director of the Guggenheim
Aeronautical Laboratory, c. 1930.
Courtesy of the Archives, California
Institute of Technology.

19 Staff of the Guggenheim
Aeronautical Laboratory in a Christ-
mas greeting at the end of Kármán's
first year as director. Courtesy of the
Archives, California Institute of
Technology.

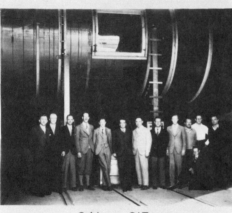

20 Men of the Austro-Hungarian Air
Force, c. 1917. The bench on which
they are seated is one of the dual
counterrotating propellers from the
tethered helicopter developed by
Kármán and others. Courtesy of the
Archives, California Institute of
Technology.

21 Kármán (forward seat) and Anthony Fokker, 1921. Courtesy of the Archives, California Institute of Technology.

22 World record flight of Aachen's
Schwarzer Düwel glider on the Wasser-
kuppe, August 30, 1921. Karman
served as advisor to the Aachen glider
team. Courtesy of the Archives,
California Institute of Technology.

23 First flight of the jet-assisted Er-
coupe, August 12, 1941. National Air
and Space Museum files.

24 Aerodynamics as science. Kármán
at Princeton, 1953. Courtesy of the
Archives, California Institute of
Technology.

11

CONCLUSION

I have said little about what created the genius of Ludwig Prandtl and Theodore von Kármán. Certainly some of its cause lay in the fertile soil of Göttingen. Some lay also, in Prandtl's case, in the training in applied mechanics he had received from the eminent August Föppl of Munich. Some lay, in Kármán's case, in the training he received from Prandtl, in his family values, and in the Hungarian intellectual background out of which other noted physicists and mathematicians emerged as well.[1] But these and other early environmental factors can account only in part for their successes; they cannot explain individual brilliance. Only insofar as the works of Prandtl and Kármán derived from or conditioned their environment circa 1926 have I sought out their creativity.

Genius aside, the main point is that material and intellectual conditions, as well as quirks of personality, played an important part in founding the science of theoretical aerodynamics and the major role in seating it in America. Circumstances directed its growth, fostering its practitioners here, inhibiting them there, sometimes halting them altogether.

America was strengthening its academic system, and its science especially, at a moment in the 1920s when the German educational systems were experiencing upheaval, in both material support and political agitation. Surely Germany returned to a modest prosperity in the late 1920s, but problems introduced in the lean years continued to mount against Kármán at Aachen. He had resolved to advance Aachen's status, and though he won small gains, such as more students, support through consulting, and an addition to his institute, they counted ultimately toward driving him to the United States for release from his heavy workload and for a gain in resources. Other American institutions had hired Germans, and so Kármán was not a lone pioneer. He was an early member of the illustrious company of continental immigrants who, because of their excellence, attracted others in a process of accretion. Caltech's reputation was world caliber by 1930, for it had secured a superb faculty, including several European immigrants in the sciences.

Kármán's family and friendships contributed to both his hesitation toward and his ultimate acceptance of Caltech. It was his mother and sister who first discouraged him from leaving Europe but Epstein, Millikan, and Guggenheim whose entreaties and support finally convinced him. The efforts of Epstein, especially, in a friendly, chatty correspondence over the years 1926–1930 showed Kármán that in America there could exist warm professional relations of the sort he recalled sharing in better times in Germany. Kármán, an adaptable and usually gracious man, developed a friendship with Clark Millikan, and this also molded his favorable outlook toward the institute. In a way, young Millikan represented to Kármán the fresh approach of American science. He was, after all, only a graduate student when he played a major part in designing the Caltech wind-tunnel laboratory—an unheard of responsibility in Europe for such a junior person. Kármán quickly accepted Millikan's de facto status of importance, indeed embraced the situation enthusiastically, as his correspondence with young Millikan reveals, and this flexible attitude contributed to the ultimate success of his relocating.

Kármán's personal ambition worked to the benefit of American aerodynamics. Guggenheim's flattery and solicitousness and Millikan's and Durand's interest appealed to Kármán's self-esteem at the same time that obstacles in Germany impelled him toward the new American institute. Kármán sought almost ruthlessly to convert his reputation to greater prestige for himself and better support for his research and his family than he had enjoyed at Aachen, and Pasadena offered a good deal. Beyond this still, the special initial relationship he had set with Caltech determined that he would slide into the job in Pasadena before taking any other.

The resemblences between Caltech and Göttingen, which perhaps Kármán did not consciously observe, nevertheless attracted him. As he admitted, he relished dealing with powerful industrialists and military men. At Caltech as at Prandtl's Göttingen, funding would be secured by designing and testing airplanes as much as from university sources. And there Kármán directed his own large experimental laboratory of far greater resources than the one he left, an institute of quality comparable to Göttingen's and with the promise of growth in a direction he had set from its beginning. He acknowledged Felix Klein more than once as his forefather in applied scientific research, curiously in one major instance without mention of Prandtl as his actual teacher.[2] His mixed feelings of love and hate for Göttingen and its professors were resolved at Caltech. Caltech had begun in the 1920s like Göttingen, and in aerodynamics Kármán perpetuated the resemblance, for he found it congenial despite his estrangement from Germany.

From Klein to Prandtl to Kármán, the roots of American research in theoretical aerodynamics grew out of the late nineteenth century relationship of German industry and academe. At Göttingen especially, Klein brought the government, industrialists, and professors into an amalgam of support for applied physics, particularly through several associations that fostered academic aerodynamics. In the name of benefit to the German Reich and so therefore to the industrialists who supported it, Prandtl increased and deepened research in the field in his "special technical institute" (in the phrase of Born's). The mode of its operation evolved from the model of the German pure physics institute. In it, the director, who was usually the only occupant of a full professorship in the institute, set the goals of his laboratory, and assistants carried out the research under his close scrutiny, as "house theoreticians" or experimenters. Graduate students were trainees, a cheap labor force of high quality, performing experiments as the director or assistants assigned them. Mechanics and other nonacademic persons built apparatus and maintained the facilities. Although the system was not unknown in America, it seemed to Hunsaker, Ames, and Guggenheim that there were few academic centers of aerodynamics in this country and none so large or with the same organization as in Germany. For the purposes of the aeronautical movement, the fact was not so important as their perception. Complaints of the Americans who wrote of having no "science of flight" here in the first quarter of the century were probably exaggerated, but they underscored an essential weakness that they wished to correct by recreating the German methods and institutions of academic study.

I might extend this line of thinking and suggest that resting in an academic environment was crucial to the success of the discipline. It was insufficient to bring a theorist like Max Munk to an existing center of developmental research such as the NACA Langley laboratory and expect that his effect would grow in an organic process if denied fertile soil. For, as George Lewis remarked, the usual designer of aircraft or indeed even the experimenters at Langley could not use his recondite work. Neither could they use Kármán's; but his creation of an academic institute of aerodynamics added to his work the essential environment for its self-perpetuation. His disciples would carry the discipline. Kármán prudently explained his science to nonspecialists and fostered a class of experimental investigations that were of use to the airplane industry.

Aerodynamics at Caltech meant different things to the different people from whom it gained support. Airplane builder

Donald Douglas could envision using the Caltech wind tunnel for design in 1925, and this to him was aeronautical science; so he supported Millikan's proposal to Guggenheim. Millikan, however, viewed aerodynamics as a science with technological uses that might appeal to Caltech's backers but at the same time possess legitimate academic standing; Guggenheim invested in aeronautical education as a gilt-edge security, a solid venture with promise of inexorable growth and guaranteed dividends; and Kármán found yet another meaning in placing aerodynamics at Caltech. For him it represented a new life in the new world of American aviation, not just professional achievement but a source of pleasure, a means of drawing attention to himself at its center and of gaining academic prestige and access to military and economic power. Earlier advocates of the scientific study of flight like Ames and Hunsaker also affirmed its benefits with such conviction as to create a climate favorable to bringing the new science to America.

This diversely motivated core of supporters recognized the uniqueness of Kármán's expertise. His showmanship, professionalism, solidity, and grace worked toward convincing them not only of the importance of the field but of his consummate abilities in it.

These sometimes ironic external conditions of economics, industrial ideology, personal ambition, military advantage, academic prestige, and political upheaval contributed as well to the growth in the 1920s of other sciences in this country. Indeed, my arguments have drawn heavily upon earlier studies of such conditions in pure physics in Europe and the United States. The reasons for the rise of applied sciences in this country deserve similar examination by case. They would show not mainly the broad historical trends of support or harassment of science, but in finer detail the inevitable links of pure and applied science and the technology which, in the utilitarian American culture, has motivated much of what we call "intellectual" pursuit. We may find as a complement to currently favored theories of the links of science and technology a close, quasi-causal relationship between particular sciences and technologies, carried not by industry, discipline or profession, but through institutions such as Göttingen and Caltech, and their benefactors, the foundations. Thus, certain of the developments of science and technology are in common subject to the outside events that affect their shared institutions.

NOTES

UNPUBLISHED SOURCES
I have used the following abbreviations in notes to designate collections of unpublished papers. In the Caltech collections, "Kármán, Box 4.22," for example, refers to Box 4, File 22, of the Theodore von Kármán Collection. When files are not numbered, either they are named or only boxes are listed.

Collections
Bayerisches Hauptstaatsarchiv:
Akten des K. Staatsministeriums des Innern für Kirchen- und Schul-Angelegenheiten, Abteilung I, Allgemeines Hauptstaatsarchiv, Bayerisches Hauptstaatsarchiv, Munich, Germany.

Brown:
Chancellor Elmer E. Brown Papers, New York University Archives, New York City.

Deutsches Museum:
Sondersammlungen der Bibliothek des Deutschen Museums, Munich, Germany.

Guggenheim Fund:
The papers of the Daniel Guggenheim Fund for the Promotion of Aeronautics, Manuscript Division, Library of Congress, Washington, D.C. In citations of correspondence, "Guggenheim" is Harry F. Guggenheim.

Hoover Commerce:
The Commerce Papers at the Herbert Hoover Presidential Library, West Branch, Iowa.

Kármán:
The Theodore von Kármán Collection at the California Institute of Technology, Archives, Pasadena, California.

Nachlass Klein:
Klein papers in the Handschriften-Abteilung der Niedersächsishen Staats- und Universitätsbibliothek, Göttingen, Germany.

Millikan:
The Robert Andrews Millikan Collection at the California Institute of Technology, Archives, Pasadena, California. In citations of correspondence, "Millikan" is Robert A. Millikan.

Mises:
The Papers of Richard von Mises, Harvard University Archives, Nathan Marsh Pusey Library, Harvard University, Cambridge, Massachusetts.

Morgan:
The Papers of Thomas Hunt Morgan at the California Institute of Technology, Archives, Pasadena, California.

T. H. Munich:
"Akten betreffend Besetzung der Lehrstellen durch Professoren-Berufungsverhandlungen, von 1922 bis 1932, Registrar-Abteilung II la. Band I," bound files of Registratur der Technischen Universität München, Munich, Germany.

NACA:
The Papers of the National Advisory Committee for Aeronautics, Record Group 255, National Archives, Washington, D.C.

NYU Archives:
New York University Archives, Washington Square, New York City.

Scherer:
The Papers of J. Scherer at the California Institute of Technology, Archives, Pasadena, California.

Wright:
The Papers of Wilbur and Orville Wright, Manuscript Division, Library of Congress, Washington, D.C.

PREFACE
1 Theodore von Kármán with Lee Edson, *The Wind and Beyond: Theodore von Kármán, Pioneer in Aviation and Pathfinder in Space* (Boston: Little, Brown, 1967) gives a charming if not always accurate account of Kármán's actions and influence. Notable reviews of his achievements are Charles Süsskind, "Theodore von Kármán," *Dictionary of Scientific Biography*, vol. VII (New York: Scribner's, 1973), pp. 246–48; Hugh L. Dryden, "Theodore von Kármán," *Biographical Memoirs of the National Academy of Sciences*, vol. 38 (New York: Columbia University Press, 1965), pp. 345–84; Dryden, "The Contributions of Theodore von Kármán: A Review," *Aeronautics and Aerospace Engineering*

(July 1963): 12–17; and Dryden, "The Contributions of Theodore von Kármán to Science and Technology," in *Collected Works of Theodore von Kármán*, vol. I (London: Butterworths Scientific Publications, 1956), pp. vii–ix, among others.

Many testimonials to Kármán's influence are cited in R. Cargill Hall's short biography, "Shaping the Course of Aeronautics, Rocketry, and Astronautics, Theodore von Kármán, 1881–1963," *The Journal of the Astronautical Sciences* 26 (1978):369–86.

2 Paul Forman, John L. Heilbron, and Spencer Weart, "Physics *circa* 1900: Personnel, Funding, and Productivity of the Academic Establishments," *Historical Studies in the Physical Sciences* 5 (1975): esp. pp. 115–28; and Daniel J. Kevles, *The Physicists: The History of a Scientific Community in Modern America* (New York: Knopf, 1978), pp. 60–90. See also Charles Weiner, "A New Site for the Seminar: The Refugees and American Physics in the Thirties," in *The Intellectual Migration: Europe and America, 1930–1960*, ed. Donald Fleming and Bernard Bailyn (Cambridge, Mass.: Harvard Belknap, 1969), pp. 190–234, esp. pp. 213, 225, and see editors' introduction, p. 8.

CHAPTER 1
1 William Glenn Cunningham, *The Aircraft Industry: A Study in Industrial Location* (Los Angeles: L. L. Morrison, 1951), p. 38.

2 Herman O. Stekler, *The Structure and Performance of the Aerospace Industry* (Berkeley: University of California Press, 1965), p. 4. Data adapted from *Aerospace Facts and Figures* (1962).

3 Howard Mingos, "The Rise of the Aircraft Industry," in *The History of the American Aircraft Industry: An Anthol-*

ogy, ed. Gene R. Simonson (Cambridge, Mass.: MIT Press, 1968), pp. 23–69, esp. pp. 22–29, 53. This essay first appeared as part of Mingos, *Birth of an Industry* (New York: W.B. Conkey, 1930), pp. 10–95.

4 Mingos, "The Rise of the Aircraft Industry," p. 54.

5 Stekler, *The Structure and Performance of the Aerospace Industry*, p. 4.

6 Quoted in Mingos, "The Rise of the Aircraft Industry," p. 59.

7 Richard P. Hallion, *Legacy of Flight: The Guggenheim Contribution to American Aviation* (Seattle: University of Washington Press, 1977), pp. 11, 14.

8 Mingos, "The Rise of the Aircraft Industry," p. 59.

9 Douglas is quoted by *Los Angeles Times* in Cunningham, *The Aircraft Industry*, p. 39.

10 Statement "Courses in Aeronautical Engineering and Industrial Aviation in the Department of Mechanical Engineering, New York University" [c. early 1923], Brown, file "Bliss, Professor Collins P., 1919/23"; and "The Quentin Roosevelt Memorial Chair of Aeronautics," fund-raising prospectus, 1925, Brown, file "Bliss, Professor Collins P." [n. d.].

11 Cunningham, *The Aircraft Industry*, pp. 39–40; Mingos, "The Rise of the Aircraft Industry," p. 53.

12 Address of Chancellor Brown, October 25, 1925, typescript, Brown, file "Guggenheim School of Aeronautics 1926/27."

13 Hallion, *Legacy of Flight*, pp. 26, 56–57.

14 Ibid., p. 58; Walter G. Vincenti, "The Air-Propeller Tests of W. F. Durand and E. P. Lesley: A Case Study in Technological Methodology," *Technology and Culture* 20 (1979):712–51.

15 Reginald M. Cleveland, *America Fledges Wings: The History of the Daniel Guggenheim Fund for the Promotion of Aeronautics* (New York: Pitman, 1942), p. 155.

16 Ibid., pp. 137–44.

17 *Reports of Officers*, 1910–, NYU Archives; and Charles H. Snow "Report of the Dean of the College of Engineering," 1922–23, New York University, New York, December 1923, pp. 139–40.

18 Charles H. Snow, "Report of the Dean of the College of Engineering," 1923–24, New York University, New York, December 1924, p. 151.

19 Hallion, *Legacy of Flight*, pp. 28–29.

20 New York University *College of Engineering* (Bulletin Series) 24, no. 3 (February 1924): 31.

21 New York University *College of Engineering, Announcements for the Year 1926–27*, pp. 68–69; Collins Bliss to Elmer E. Brown, January 30, 1924, Brown, file, "Bliss, Professor Collins P." [n.d.]; and Alexander Klemin, "Aviation and the University," *The Scientific Monthly* 23 (September 1926): 284.

22 *Daniel Guggenheim School of Aeronautics, New York University College of Engineering, 30th Anniversary*, New York University, 1955?, p. 17; Alexander Klemin, "Brief History of the Daniel

Guggenheim Graduate School of Aeronautics of the College of Engineering, NYU," p. 6, draft, with note attached from F. Teichmann dated October 29, 1948, NYU Archives.

23 Eduard C. Lindeman, *Wealth and Culture* (New York: Harcourt Brace, 1936), p. 15.

24 Ibid., p. 26.

25 Kevles, *The Physicists*, pp. 148–49, 190–91; David F. Noble, *America by Design: Science, Technology, and the Rise of Corporate Capitalism* (New York: Knopf, 1977), pp. 126–30.

26 Noble, *America by Design*, pp. 129–30.

CHAPTER 2
1 Millikan to Carty, May 17, 1926, Millikan, Box 16.6.

2 Epstein to Kármán, July 5, 1926, Kármán, Box 8.

3 Robert Kargon points to this technological expansion as characteristic of the era in his forthcoming biography of Robert Millikan. The phenomenon, Millikan's central position in it, and one interpretation of its causes are discussed in David F. Noble, *America by Design: Science, Technology, and the Rise of Corporate Capitalism* (New York: Knopf, 1977), esp. pp. 152–55.

4 Chandler to Millikan, December 24, 1925; Douglas to Millikan, December 24, 1925; Robert A. Morton to Millikan, December 28, 1925; Millikan, Box 16.5.

5 Richard P. Hallion, *Legacy of Flight: The Guggenheim Contribution to America Aviation* (Seattle: University of Washington Press, 1977), p. 51. As Hallion notes, general terms of the grant had been published in a "Bulletin of the Daniel Guggenheim Fund for the Promotion of Aeronautics, Inc.," no. 1, August 14, 1926 (Guggenheim Fund, Box 6). Details and announcement of Kármán's visit appeared September 6 in "Guggenheim Fund Endows Universities," *Aviation* 21 (1926):402.

6 Millikan to Kármán, telegram, July 26, 1926, Millikan, Box 16.6.

7 Millikan to F. B. Jewett, January 18, 1926, Millikan, Box 16.6.

8 Hallion, *Legacy of Flight*, pp. 37–38.

9 On the relations of Carty, Jewett, and Millikan, see Daniel J. Kevles, *The Physicists: The History of a Scientific Community in Modern America* (New York: Knopf, 1978), p. 99; and Noobar R. Danielian, *AT&T: The Story of Industrial Conquest* (New York: Vanguard Press, 1939), pp. 104–105.

10 Millikan to Merriam, January 18, 1926, Millikan, Box 16.6.

11 Theodore von Kármán, *Aerodynamics* (New York: McGraw-Hill, 1954), p. 12.

12 Proposal attached to letter, Millikan to H. Guggenheim, December 24, 1925, Guggenheim Fund, Box 6. Note attached, dated January 7, 1926, says papers were handed in person to H. F. Guggenheim and matter "was discussed fully" the day before.

13 Millikan to Guggenheim, January 29, 1926, Guggenheim Fund, Box 6.

14 H. I. Cone to Millikan, February 8, 1926, Guggenheim Fund, Box 6.

15 On Kármán's involvement in patent disputes, see Theodore von Kármán with Lee Edson, *The Wind and Beyond: Theodore von Kármán, Pioneer in Aviation and Pathfinder in Space* (Boston: Little Brown, 1967), pp. 111–15. On allied suspicion and German concern for representing aeronautical research as nonmilitary, see p. 120. NACA representatives encountered no problems in gaining access to Prandtl's laboratory, in part because they supported his work financially. See Reports on European Aviation, 1920–1923, NACA, Series 18, Box 1.

16 Prandtl's role in patent disputes is suggested in Kármán to Prandtl, July 19, 1930; and Prandtl to Kármán, July 23, 1930; Kármán, Box 23.

17 Guggenheim, "Report," Hoover Commerce, p. 4.

18 Guggenheim to Millikan, telegram, April 28, 1926, Guggenheim Fund, Box 6.

19 Millikan to Guggenheim, May 14, 1926, Guggenheim Fund, Box 6.

20 Robert H. Kargon, "Temple to Science: Cooperative Research and the Birth of the California Institute of Technology," *Historical Studies in the Physical Sciences* 8 (1977): 22; and "Plan for an Institute of Physics and Chemistry," Scherer, Box 2.

21 "Minutes of Special Meeting of the Board of Directors," June 2, 1926, Wright, Box 21; Guggenheim to Millikan, June 7, 1926, Guggenheim Fund, Box 6.

22 "Letter Report to the Members of the Daniel Guggenheim Fund," June 2, 1926, Wright, Box 21.

23 Millikan to Carty, May 17, 1926, Millikan, Box 16.6.

24 Michelson to Millikan, June 3, 1926, with copy of proposed draft, Michelson to Millikan, Millikan, Box 16.6.

25 "Guggenheim Fund Makes Grants to Finance Study and Experiments in Aeronautics," Bulletin of the Daniel Guggenheim Fund, no. 1, Guggenheim Fund, Box 6.

26 For another view of Kármán's part in aeronautics in America, see Reginald M. Cleveland's authorized *America Fledges Wings: The History of the Daniel Guggenheim Fund for the Promotion of Aeronautics* (New York: Pitman, 1942), pp. 137–44.

27 Millikan, copy of announcement in Pasadena, August 8, 1926, Guggenheim Fund, Box 6.

28 Karl-Heinz Manegold, "Universität, Technische Hochschule und Industrie: Ein Beitrag zur Emanzipation der Technik im 19. Jahrhundert unter besonderer Berücksichtigung der Bestrebungen Felix Kleins," *Schriften zur Wirtschafts- und Sozialgeschichte* 16 (Berlin: Duncker and Humblot, 1970): pp. 167–75; and Lewis Pyenson, "Einstein's Early Scientific Collaboration," *Historical Studies in the Physical Sciences* 7 (1976): 84–88.

CHAPTER 3
1 Daniel J. Kevles, *The Physicists: The History of a Scientific Community in Modern America* (New York: Knopf, 1978), p. 44.

2 Russell McCormmach, Editor's Foreword, *Historical Studies in the Physical Sciences* 3 (1971):xv.

3 Paul Forman, John L. Heilbron, and Spencer Weart, "Physics *circa* 1900: Personnel, Funding, and Productivity of the Academic Establishments," *Historical Studies in the Physical Sciences* 5 (1975):80–81, 102–03; and McCormmach, Editor's Foreword, pp. xv–xvii.

4 Karl-Heinz Manegold, "Universität, Technische Hochschule und Industrie: Ein Beitrag zur Emanzipation der Technik im 19. Jahrhundert unter besonderer Berücksichtigung der Bestrebungen Felix Kleins," *Schriften zur Wirtschafts- und Sozialgeschichte* 16 (Berlin; Duncker and Humblot, 1970): 158; Werner Burau and Bruno Schoeneberg, "Christian Felix Klein," *Dictionary of Scientific Biography*, vol. VII (New York: Scribner's, 1973), p. 399; and W. Tollmien, "Zum Geleit," in *Ludwig Prandtl Gesammelte Abhandlungen zur angewandten Mechanik, Hydro and Aerodynamic*, vol. I, ed. W. Tollmien, H. Schlichting, and Henry Görtler (Berlin: Springer, 1961), p. v.

5 Constance Reid, *Courant in Göttingen and New York: The Story of an Improbable Mathematician* (New York: Springer-Verlag, 1976), pp. 101, 117–18.

6 Burau and Schoeneberg, "Klein," p. 399.

7 Edwin T. Layton, Jr., "American Ideologies of Science and Engineering," *Technology and Culture* 17 (1976): 688–700.

8 Felix Klein to Carl Runge, July 13, 1896, Deutsches Museum, Klein-Runge Briefwechsel, file "1950/6."

9 Paul Forman, "Carl David Tolmé Runge," *Dictionary of Scientific Biography*, vol. XI (New York: Scribner's, 1975), pp. 610–15.

10 Burau and Schoeneberg, "Klein," p. 399.

11 For a discussion of the seal, see K. Kraemer, "Geschichte der Gründung des Max-Planck-Instituts für Strömungsforschung in Göttingen," in *Max-Planck-Institut für Strömungsforschung Göttingen 1925–1975*: Festschrift zum 50jährigen Bestehen des Instituts (Göttingen: MPI für Strömungsforschung Göttingen, 1975), pp. 17–19.

12 Kenneth O. May, "Carl Friedrich Gauss," *Dictionary of Scientific Biography*, vol. V (New York: Scribner's 1972), pp. 298–315.

13 A. E. Woodruff, "Wilhelm Eduard Weber," *Dictionary of Scientific Biography*, vol. XIV (New York: Scribner's, 1976), pp. 203–209.

14 McCormmach, Editor's Foreword, p. xiii.

15 The Department of Applied Physics, as Kármán recalled it, was called sarcastically "the Lubricating Oil Faculty," a phrase that reflected also its hydraulic studies of petroleum. Theodore von Kármán with Lee Edson, *The Wind and Beyond: Theodore von Kármán, Pioneer in Aviation and Pathfinder in Space* (Boston: Little, Brown, 1967), p. 35.

16 Max Born, *My Life: Recollections of a Nobel Laureate* (New York: Scribner's, 1978), pp. 208–9.

17 Richard von Mises, "Felix Klein. Zu seinem 75. Geburtstag am 25. April 1924," *Zeitschrift für angewandte Mathematik und Mechanik* 4 (1924): 86–92, esp. p. 91.

18 Ibid.

19 Klein's energy and sophistication are amply evident in hundreds of letters and manuscripts preserved among his papers at Göttingen, to which I return below.

20 Lewis Pyenson, "Einstein's Early Scientific Collaboration," *Historical Studies in the Physical Sciences* 7 (1976): 88.

21 Reid, *Courant*, p. 26.

22 Pyenson, "Einstein's Early Scientific Collaboration," pp. 85–88. See also Pyenson's "Goettingen Reception of Einstein's General Theory of Relativity," Ph.D. dissertation, The Johns Hopkins University, 1973, pp. 156–88.

23 Pyenson, "Einstein's Early Scientific Collaboration," p. 86; and Reid, *Courant*, pp. 46, 84.

24 Manegold, "Universität, Technische Hochschule und Industrie," p. 159. One example: the Aachen Polytechnic grew in enrollment from about 200 in 1890 to 500 in 1900. The following account of Klein's gathering of support follows Manegold, pp. 159–75.

25 File of letters between Klein and A. Cayley, D. C. Gilman, and Paul Haupt, "Briefe u. Akten betr. Berufung nach Baltimore," Nachlass Klein, file 22L:7.

26 Manegold, "Universität, Technische Hochschule und Industrie," p. 158.

27 Friedrich Klemm, "Carl von Linde," *Dictionary of Scientific Biography*, vol. VIII (New York: Scribner's, 1973), pp. 365–66.

28 Manegold, "Universität, Technische Hochschule und Industrie," p. 165.

29 Paul Forman, *The Helmholtz-Gesellschaft: Support of Academic Physical Research by German Industry after the First World War*, Smithsonian Institution, unpublished monograph, 1971, p. 6.

30 Manegold, "Universität, Technische Hochschule und Industrie," p. 173.

31 Rieppel's views were paid heed by the academic founders of the association, even if there was an uneasy tension. In 1912, apparently because of waning interest in the essentially academic association, Rieppel sought to resign his company's membership. Böttinger quickly wrote Klein, observing that Rieppel's leaving would be dangerous for the stability of the group because it might precipitate a wave of resignations of other industrialists representing firms that are "only half committed to the thing" anyway. He recommended that Klein write Rieppel, which evidently Klein did with success, for Rieppel remained. See especially Böttinger to Klein, March 2, 1912, Nachlass Klein, file 4F "Göttinger Vereinigung 1908–1912."

CHAPTER 4
1 Prandtl used the term in a letter to Klein, May 9, 1904, Nachlass Klein, file 2F "Neuberufung 1904." Also Lewis Robert Pyenson, "The Goettingen Reception of Einstein's General Theory of Relativity," Ph.D. dissertation, The Johns Hopkins University, 1973, p. 130.

2 Naumann to Klein, June 7, 1904, Nachlass Klein, file 1D "Amtliches 1900–1909."

3 Klein to Althoff, May 21, 1904, and Böttinger to Klein, June 16, 1904, Nachlass Klein, file 2F "Neuberufung 1904."

4 Stodola recommended Prandtl to Klein in 1900, citing his "far beyond the average gift" of mathematical ability. Stodola to Klein, May 13, 1900; also A. Föppl to Klein, May 17, 1900, and A. Rieppel to Klein, May 16, 1900, all letters of recommendation solicited by Klein, Nachlass Klein, file 2F "Neuberufung 1904."

5 Klein to Elster, June 27, 1904, Nachlass Klein, file 1D "Amtliches 1900–1909."

6 Paul Forman, "Carl David Tolmé Runge," *Dictionary of Scientific Biography*, vol. xi, (New York: Scribner's 1975), p. 613; and "Biographische Daten und Ehrungen," *Ludwig Prandtl Gesammelte Abhandlungen* . . ., III, p. 1619. Hereafter *Prandtl Abhandlungen*.

7 Prandtl to Klein, June 25, 1904, Nachlass Klein, file 2F "Neuberufung 1904."

8 Prandtl to Klein, June 27, 1904, Nachlass Klein, file 2F "Neuberufung," and Runge's draft response on Klein to Runge, July 17, 1904, Deutsches Museum, Klein–Runge Briefwechsel, file "1950/6."

9 Prandtl to Klein, June 25, 1904, Nachlass Klein, file 2F "Neuberufung."

10 Prandtl to Klein, May 9, 1904, Nachlass Klein, file 2F "Neuberufung 1904."

11 Naumann to Klein, June 9, 1904, Nachlass Klein, file 1D "Amtliches 1900–1909."

12 *Die Physikalischen Institute der Universität Göttingen. Festschrift*, published by the Göttinger Vereinigung zur Forderung der angewandten Physik und Mathematik (Leipzig: B. G. Teubner, 1906), pp. 95–111; and "Tagesereignisse," *Physikalische Zeitschrift* 7 (1906):72.

13 Ludwid Prandtl, "Felix Klein zum Gedächtnis," *Zeitschrift für Flugtechnik und Motorluftschiffahrt* 16 (1925):275.

14 "Ludwig Prandtl," citation from *The Daniel Guggenheim Medal for Achievement in Aeronautics* (New York: n.p., 1936), p. 14.

15 Felix Klein, "Aeronautics at Göttingen University: Some Fields of Research Which Are to be Explored," *Scientific American Supplement* 67 (1909):287. On Prandtl's promotion, see "Biographische Daten und Ehrungen," *Prandtl Abhandlungen* III: 1619.

16 Böttinger affirmed Prandtl's sway with the gentlemen of the Kaiser Wilhelm Society, which supported his institute, in Böttinger to Klein, June 18, 1913, Nachlass Klein, file 4C "Institut für Aerodynamik [1912–1921] Briefe und Akten."

17 Klein, "Aeronautics at Göttingen"; and "University and Educational News," *Science* 29 (1909):613.

18 E. Ehrensberger to Klein, October 28, 1908, Nachlass Klein, file 4F "Göttinger Vereinigung 1908–1912." Ehrensberger was Director of the Krupp firm.

19 Ehrensberger to Klein, November 9, 1908, Nachlass Klein file 4F "Göttinger Vereinigung 1908–1912." Announcement of Krupp's award was made simultaneously with that of Krupp's election to membership in the association on November 21.

20 "Ludwig Prandtl," p. 14.

21 Klein, "Aeronautics at Göttingen."

22 Prandtl, "Felix Klein."

23 Böttinger to Klein, November 9, 1911, and March 22, 1912, Nachlass Klein, file 7C "Luftfahrt 1906–1912," show reservations on Böttinger's part and then wholehearted collaboration with Prandtl to secure the new institute. Böttinger to members of the Göttingen Association, June 23, 1913, announces KW Society approval. Nachlass Klein, file 4C "Institut für Aerodynamik [1912–1921] Briefe und Akten."

24 Böttinger to Prince Heinrich of Prussia, May 18, 1915, Nachlass Klein, file 4C "Institut für Aerodynamik."

25 Böttinger to Klein, March 29, 1915, Nachlass Klein, file 3J "Bottingensia. Laufendes 1913–1915." Also Böttinger's report to the members of the Göttingen Association, May 21, 1915, and Prandtl's proposal, "Ausbau der Göttinger Modellversuchsanstalt zu einem vollwertigen aerodynamischen Forschungsinstitut für Herr und Marine," April 26, 1915, Nachlass Klein, file 4C "Institut für Aerodynamik."

26 Böttinger to Klein, May 8, 1915, Nachlass Klein, file 4C "Institut für Aerodynamik."

27 Böttinger to Prince Heinrich of Prussia, May 18, 1915, Nachlass Klein, file 4C "Institut für Aerodynamik," and John Howard Morrow, Jr., *Building German Airpower, 1909–1914* (Knoxville: University of Tennessee Press, 1976), pp. 58, 117.

28 "The Goettingen Aerodynamical Laboratory," *Aeronautical Engineering, Supplement to "The Aeroplane"* (August 2, 1922): 87–89; and L. Prandtl, ed., *Ergebnisse der Aerodynamischen Versuchsanstalt zu Göttingen*, vol. I (Munich: R. Oldenbourg, 1921), pp. 1–7.

29 Constance Reid, *Courant in Göttingen and New York: The Story of an Improbable Mathematician* (New York: Springer-Verlag, 1976), pp. 42, 83.

30 "Aerodynamik," *Encyklopädie der mathematischen Wissenschaften mit Einschluss ihrer Anwendungen*, vol. IV, no. 17 (Leipzig: Teubner, 1902), 149–84.

31 William F. Durand, ed., *Aerodynamic Theory: A General Review of Progress* (Berlin: J. Springer, 1934–36).

32 Mises, handwritten introduction to lecture by K. Popoff, May 17, 1926, Mises.

33 Theodore von Kármán with Lee Edson *The Wind and Beyond*, (Boston: Little, Brown, 1967), pp. 37–40; and Reid, *Courant*, p. 145.

CHAPTER 5
1 Kármán, "Ludwig Prandtl," *Zeitschrift für Flugtechnik und Motorluftschiffahrt* 16 (1925):37.

2 Hunter Rouse, *Hydraulics in the United States, 1776–1976* (Iowa City:

Institute of Hydraulic Research, 1976), pp. 125–26.

3 L. Prandtl, "Über Flüssigkeits-bewegung bei sehr kleiner Reibung," *Verhandlungen des III. Internationalen Mathematiker-Kongresses, Heidelberg 1904*, ed. A. Krazer (Leipzig: Teubner, 1905), pp. 484–91; re-printed in L. Prandtl and A. Betz, *Vier Abhandlungen zur Hydrodynamik und Aerodynamik* (Göttingen: Selbst-verlag der AVA, 1927, reissued 1944), pp. 1–8. Future references are to this reprint. Also *Prandtl Abhandlungen* II:575–84.

4 Hunter Rouse and Simon Ince, *History of Hydraulics* (Iowa City: Institute of Hydraulic Research, 1957), p. 230.

5 "Über Flüssigkeitsbewegung," p. 2; English translation from Rouse and Ince, p. 230. The 1904 paper was earlier translated into English and published as NACA Technical Memo-randum no. 452, Washington, D.C., 1928. Rouse and Ince have performed their own partial translation, which is more accurate than that of the NACA.

6 "Über Flüssigkeitsbewegung," p. 3, my translation.

7 "Beitrag zur näherungsweisen Inte-gration totaler Differentialgleichung-en," *Zeitschrift für Mathematik und Physik* 46 (1901):435–53.

8 Theodore von Kármán, *Aerodynamics* (New York: McGraw-Hill, 1954), p. 12; and A. F. Zahm, "Theoretical Aerodynamics," *Mechanical Engineering* 52 (1930):499–500.

9 "Über Flüssigkeitsbewegung," pp. 4–5.

10 Ibid., p. 6. This is my translation and differs somewhat from Rouse and Ince's, p. 230.

11 Nachlass Klein, files 7C "Luftfahrt 1906–1912," and 4F "Göttinger Ver-einigung 1908–1912."

12 Theodore von Kármán with Lee Edson, *The Wind and Beyond* (Boston: Little, Brown, 1967): 61.

13 Gerald Holton, "Influences on Ein-stein's Early Work," in his *Thematic Origins of Scientific Thought: Kepler to Einstein* (Cambridge, Mass.: Harvard University Press, 1973), p. 207.

14 *Vorlesungen über technische Mechanik*, 1898, cited in Holton, p. 207.

15 Kármán, *The Wind and Beyond*, p. 40.

16 Kármán, "Ludwig Prandtl," pp. 37–40.

17 "Ludwig Prandtl," citation from *The Daniel Guggenheim Medal for Achievement in Aeronautics* (New York: n.p., 1936), pp. 14–16.

18 "Tragflügeltheorie: I. Mitteilung," *Nachrichten der Königlichen Gesellschaft der Wissenschaften zu Göttingen*, Mathematisch-physikalische Klasse (hereafter *Göttinger Nachrichten*) (1918):451–77; and "Tragflügeltheorie: II. Mitteilung," *Göttinger Nachrichten* (1919):107–37; reprinted in Prandtl and Betz, *Vier Abhandlungen*, pp. 7–67. For a cursory overview of Prandtl's wing theory, see also John H. Lienhard, "Ludwig Prandtl," *Dictionary of Scientific Biography*, vol. XI (New York: Scribner's, 1975), pp. 123–125.

19 Kármán, "Ludwig Prandtl," p. 38.

20 L. Prandtl, ed. *Ergebnisse der Aerodynamischen Versuchsanstalt zu Göttingen*, vol. I (Munich: R. Oldenbourg, 1921), p. 4.

21 Ibid. After the revolution of 1918 staff and financing dwindled, as Prandtl notes also. The next three volumes of these *Ergebnisse* show that prosperity returned to the Göttingen laboratory in the next few years, however.

22 Rouse and Ince, *History of Hydraulics*, pp. 234–35.

23 Kármán, *The Wind and Beyond*, p. 41.

24 Max Born to Albert Einstein, July 15, 1925, in *The Born–Einstein Letters*, ed. Max Born (New York: Walker, 1971), p. 85. Born rued the disturbance: "There is going to be another rumpus tomorrow; it is the inauguration of Prandtl's new hydrodynamics institute, with guided tour, official dinner and gala concert. It will cost me almost an entire working day."

CHAPTER 6
1 Theodore von Kármán with Lee Edson, *The Wind and Beyond* (Boston: Little, Brown, 1967), pp. 26–34.

2 Biographical facts, except where otherwise noted, are taken from Kármán's *The Wind and Beyond*, pp. 1–146; Hugh L. Dryden, "Theodore Von Kármán," *Biographical Memoirs of the National Academy of Sciences*, vol. 38 (New York: Columbia University, 1965), pp. 345–84; Hugh L. Dryden, "The Contributions of Theodore von Kármán: A Review," *Astronautics and Aerospace Engineering* (July 1963):

12–17; Kurt Düwell, "Gründung und Entwicklung der Rheinisch-Westfälischen Technischen Hochschule Aachen bis zu ihrem Neuaufbau nach dem Zweiten Weltkrieg— Darstellung und Dokumente," in *Rheinisch-Westfälische Technische Hochschule Aachen 1870/1970*, vol I, ed. Hans Martin Klinkenberg (Stuttgart: Oscar Bek Verlag, 1970), pp. 19–175; and Charles Süsskind, "Theodore von Kármán," *Dictionary of Scientific Biography*, vol. VII (New York: Scribner's, 1973), pp. 246–248.

3 "Über den Mechanismus des Widerstandes, den ein bewegter Körper in einer Flüssigkeit erfährt. 1. Teil," *Göttinger Nachrichten* (1911): pp. 509–17; reprinted in *Collected Works of Theodore von Kármán*, vol. I (London: Butterworths, 1956), pp. 324–30.

4 Kármán, *Aerodynamics* (New York: McGraw-Hill, 1954), p. 68.

5 Reynolds, "An Experimental Investigation of the Circumstances Which Determine Whether the Motion of Water Shall Be Direct or Sinuous and of the Law of Resistance in Parallel Channels," *Philosophical Transactions of the Royal Society*, Series A, 174 (1883): 935–82.

6 "Über den Mechanismus, 1. Teil," p. 324.

7 Kármán summarized the situation, giving citations as he did not in the 1911 paper, in *Aerodynamics*, pp. 25–28. The relevant works were: d'Alembert, *Essai d'une nouvelle théorie de la résistance des fluides* (Paris, 1752); H. von Helmholtz, "Über discontinuirliche Flüssigkeitsbewegungen," *Berliner Monatsberichte* (1868):215–28; G. Kirchhoff, "Zur Theorie freier

Flüssigkeitsstrahlen," *Zeitschrift für die reine und angewandte Math.* 70 (1869):289–98; and Lord Rayleigh, "On the Resistance of Fluids," *Philosophical Magazine*, Series 5, 2 (1876):430–41.

8 "Über den Mechanismus, 1. Teil," p. 325.

9 W. Thomson, *Mathematical and Physical Papers* (Cambridge: Cambridge University Press, 1882–1911), vol. IV, p. 215.

10 Prandtl already in "Über Flüssigkeitsbewegung," pp. 6–7, commented to this effect.

11 Kármán, *Aerodynamics*, p. 70.

12 Kármán wrote the condition for stability as $Cos(\pi h/l) = \sqrt{3}$, where h is height and l is horizontal separation. The gothic "Cos" was not the pure trigonometric function of cosine, but the hyperbolic cosine, written commonly in English as "Cosh." (I owe this point to a remark by Nicholas Hoff.) Thus $\pi h/l$ must equal the angle whose hyperbolic cosine is $\sqrt{3}$, a statement equivalent to the condition expressed in the text here.

13 "Über den Mechanismus, 1. Teil," pp. 325–29.

14 Ibid., p. 327

15 Ibid., pp. 329–30.

16 Ibid., p. 330.

17 Max Abraham, a theoretical physicist, made much of this analogy in his review article "Geometrische Grundbegriffe," *Encyclopädie der Mathematischen Wissenschaften*, vol. IV,

no. 7 (Leipzig: Teubner, 1901), pp. 1–47, which Prandtl cited in "Über Flüssigkeitsbewegung," p. 3. But Horace Lamb, in his *Hydrodynamics* of 1932, 6th ed., chose not to cite the equivalent law of Biot and Savart of electromagnetics.

18 On Kármán's difficulties with the English language, see *The Wind and Beyond*, pp. 27–28.

19 Lamb, *Hydrodynamics*, 6th ed., 1st American ed. (New York: Dover, 1945), pp. 225–29, 680–81.

20 Prandtl, "Über Flüssigkeitsbewegung," pp. 1, 5.

21 Kármán and Rubach, "Über den Mechanismus des Flüssigkeits- und Luftwiderstandes," *Physikalische Zeitschrift* 13 (1912):49–59; and *Collected Works*, vol. I, pp. 339–58.

22 Über den Mechanismus, 2. Teil," *Göttinger Nachrichten* (1912):547–56; and *Collected Works*, vol. I, pp. 331–38.

23 *Aerodynamics*, p. 71.

24 Kármán to Lee Edson, November 4, 1961, Kármán, Box 115.6.

25 *Aerodynamics*, p.71.

26 "Contributions of Theodore von Kármán to Applied Mechanics," *Applied Mechanics Reviews* 16 (1963): 589–95; Düwell, "Gründung und Entwicklung der RWTH Aachen," p. 74; and Kármán, *The Wind and Beyond*, p. 57.

27 "Über Längsstabilität und Längsschwingungen von Flugzeugen," *Jahrbuch der Wissenschaftlichen Gesellschaft*

für Luftfahrt 3 (1914/15):116–38; and *Collected Works*, vol. II, pp. 1–25.

28 "Potentialströmung um gegebene Tragflächenquerschnitte," *Zeitschrift für Flugtechnik und Motorluftschiffahrt* 9 (1918):111–16; and *Collected Works*, vol. II, pp. 36–51

29 "Über laminare und turbulente Reibung," *Zeitschrift für angewandte Mathematik und Mechanik* 1 (1921): 233–52; and *Collected Works*, vol. II, pp. 70–97.

30 Prandtl to Kármán, June 14, 1921, Kármán, Box 23.

31 "Über die Oberflächenreibung von Flüssigkeiten," *Vorträge aus dem Gebiete der Hydro- und Aerodynamik, Innsbruck 1922* (Berlin: Springer, 1922), pp. 146–67; and *Collected Works*, vol. II, pp. 133–52.

32 Ibid., p. 152; quoted in Dryden, "Contributions to Applied Mechanics," p. 591.

33 "Über die Stabilität der Laminarströmung und die Theorie der Turbulenz," *Proceedings of the International Congress for Applied Mechanics, Delft 1924*; and *Collected Works*, vol. II, pp. 219–36.

34 "Die mittragende Breite," *Beiträge zur technischen Mechanik*, 1924, pp. 114–27; and *Collected Works*, vol. II, pp. 176–88.

35 Dryden, "Contributions to Applied Mechanics," p. 591.

36 Bienen and Kármán, "Zur Theorie der Luftschrauben," *Zeitschrift des Vereins Deutscher Ingenieure* 68 (1924): 1237–42, 1315–18; and *Collected Works*, vol. II, pp. 189–218.

37 "Über elastische Grenzzustände," *Verhandlungen des 2. internationalen Kongresses für technische Mechanik, Zürich 1926* (Zurich: O.Fussli, 1927(?)); and *Collected Works*, vol. II, pp. 241–52.

38 "Berechnung der Druckverteilung an Luftschiffkörpern," *Abhandlungen aus dem Aerodynamischen Institut an der Technischen Hochschule, Aachen*, 6 (1927):3–17; and *Collected Works*, vol. II, pp. 253–73.

39 Kármán, *Aerodynamics*, pp. 50–51.

40 See Chapter 9.

41 Richard von Mises, for one, recognized a source of Kármán's excellence in his interdisciplinary approach: educated by Göttingen mathematicians and physicists, but with his interest applied to technology, Kármán was "predestined" to "establish a bond between science and technology." Mises to Prandtl, January 9, 1923, Mises.

42 *The Guggenheim Aeronautical Laboratory of the California Institute of Technology: The First Twenty-Five Years* (Pasadena: California Institute of Technology, 1954), p. 7; and Kármán, *The Wind and Beyond*, p. 54.

43 Both Edward P. Warner and John J. Ide, representatives of the NACA in Europe, mentioned Aachen's problems. Warner, in an undated "Report on German Wind Tunnels and Apparatus," NACA, Series 18, Box 1, commented circa 1920 on the irregular flow of the wind tunnel at Aachen, and cited an "ear-splitting shriek totally unlike anything that I ever heard from a wind tunnel before." Ide, in a "Memorandum from the Technical Assistant in Europe, NACA," of Oc-

tober 17, 1921, remarked on "the famous shriek" and continued: "The only speed at which any serious work can be done is at about 22 meters (72 ft) per second." NACA Series 18, Box 1.

44 Kármán to Lee Edson, November 4, 1961, Kármán, Box 115.6.

45 Kármán, *The Wind and Beyond*, p. 136.

46 Kármán to Edson, November 4, 1961.

CHAPTER 7
1 Millikan, "Proposal for a Research Center in Aeronautics at the California Institute of Technology." Guggenheim Fund, Box 6.

2 Klein, "Aeronautics at Göttingen University: Some Fields of Research which are to be Explored," *Scientific American Supplement* 67 (1909):287.

3 Zahm, "On the Need for an Aeronautical Laboratory in America," *Aero Club of America Bulletin*, January 1912, p. 35; "Uses of an Aeronautical Laboratory," ibid., (February-March 1912): 15; "Eiffel's Aerodynamic Laboratory and Studies," ibid., August 1912:2–6.

4 Zahm, "On the Need for an Aeronautical Laboratory in America."

5 Zahm, "Eiffel's Aerodynamic Laboratory," p. 3.

6 Zahm, "On the Need for an Aeronautical Laboratory in America."

7 "Report on European Aeronautical Laboratories," *Smithsonian Miscellaneous Collections*, vol. 62, no. 3 (1914).

8 Hunsaker, "Europe's Facilities for Aeronautical Research—I," *Flying* (April 1914):75, 93; and "Europe's Facilities, II," May 1914, pp. 108–09.

9 Hunsaker et al., "Reports on Wind Tunnel Experiments in Aerodynamics," *Smithsonian Miscellaneous Collections*, vol. 62, no. 4 (1916).

10 Zahm, "European Aeronautical Laboratories," p. 4.

11 Nathan Reingold, "The Case of the Disappearing Laboratory," *American Quarterly* (Spring 1977):79–101. See also Daniel J. Kevles, *The Physicists: The History of a Scientific Community in Modern America* (New York: Knopf, 1978,), pp. 149, 155–6, and Robert H. Kargon, "Temple to Science: Cooperative Research and the Birth of the California Institute of Technology," *Historical Studies in the Physical Sciences* 8 (1977).

12 Hunsaker, "Europe's Facilities for Aeronautical Research," pp. 75, 108, 109.

13 Richard P. Hallion, *Legacy of Flight: The Guggenheim Contribution to American Aviation* (Seattle: University of Washington Press, 1977). p. 16.

14 Minutes of NACA Executive Committee Meeting, November 11, 1920, NACA, Series 1, Box 4, p. 9.

15 Joseph S. Ames summarized Munk's major contributions in the address "Recent Progress in the Science of Aeronautics," delivered at the Centenary Celebration of the Founding of the Franklin Institute on September 17, 1924. Manuscript copy in NACA, Series 3, Box 1.

16 "The Aerodynamic Experimental Station in Göttingen," *Engineering* 112 (1921):218–20.

17 "The Göttingen Aerodynamical Laboratory," *Aeronautical Engineering, Supplement to "The Aeroplane"* (August 2, 1922):87–89; and review of *Ergebnisse der Aerodynamischen Versuchsanstalt zu Göttingen*, ed. L. Prandtl, C. Wieselsberger, and A. Betz, vol. I (1921) and vol. II (1923) in *Journal of the Royal Aeronautical Society*. (1924): 268.

18 "The Goettingen Aerodynamical Laboratory," *Aviation* 12 (1922):310.

19 Alexander Klemin, "Work of the Goettingen Aerodynamic Laboratory," *Aviation* 16 (1924):619–20.

20 "Aerodynamic Characteristics of Airfoils," vol. I, NACA Report no. 93 (1920); vol. II, NACA Report no. 124 (1921); and vol. III, NACA Report no. 182 (1924).

21 J. C. Hunsaker to Jospeh S. Ames, September 1, 1920, NACA; R. Giacomelli, "Historical Sketch," in *Aerodynamic Theory*, vol. I, ed. W. F. Durand, p. 391. The work was published as NACA Report no. 116 (1921).

22 Ames, "Aeronautic Research," *Journal of the Franklin Institute* (January 1922):15–28; reprinted in *Smithsonian Report for 1922* (Washington, D.C.: Government Printing Office, 1924), pp. 167–74.

23 Ibid., p. 15.

24 Kevles, *The Physicists*, pp. 91–184.

25 Ames, "Aeronautic Research," pp. 16, 21.

26 Ames, "Recent Progress in the Science of Aeronautics," NACA, Series 3, Box 1, p. 12.

27 Ames to Hunsaker, April 30, 1924, NACA, Series 3, Box 12.

28 L. Bairstow, "Skin Friction," *Journal of the Royal Aeronautical Society* 29 (1925):3–14, and "Discussion," ibid., 15–23.

29 Lewis to Ames, July 2, 1924, NACA, Series 3, Box 1.

30 Recollections of Fred Weick, 1977, National Air and Space Museum audio tape collection.

31 Hallion, *Legacy of Flight*, pp. 189–98.

32 Walter Vincenti, "The Air-Propeller Tests of W. F. Durand and E. P. Lesley: A Case Study in Technological Methodology," *Technology and Culture* 20 (1979):712–51.

CHAPTER 8
1 Robert H. Kargon, "Temple to Science: Cooperative Research and the Birth of the California Institute of Technology," *Historical Studies in the Physical Sciences* 8 (1977):22.

2 Ibid., quoted from "Plan for an Institute of Physics and Chemistry," Scherer, Box 2.

3 Millikan to Guggenheim, July 7, 1926, Millikan, Box 16.6.

4 Epstein recalled fifteen years later, "When Dr. Millikan called on me . . ., there was not the slightest doubt in my mind as to who the best specialist was." Handwritten manuscript of talk, probably on occasion of Kármán's six-

tieth birthday (1941), Kármán, Box 8.21.

5 Prandtl had refused once to leave Göttingen, in 1923, and later evidence shows that he worked, if reluctantly, for the German military in World War II: See Kármán, "Ludwig Prandtl," *Zeitschrift für Flugtechnik und Motorluftschiffahrt*, p. 37; on his growing association with the military before World War II, see Leslie E. Simon, *German Research in World War I* (New York: Wiley, 1947), pp. 73, 75, 76.

6 Millikan to Thomas Hunt Morgan, October 19, 1927, Morgan, Box 2; copy in Millikan, Box 27.9.

7 Epstein to Kármán, July 5, 1926, Kármán, Box 8.

8 "Salary Records: 1927–31," California Institute of Technology Archives.

9 Theodore von Kármán with Lee Edson, *The Wind and Beyond* (Boston: Little, Brown, 1967), p. 175.

10 Guggenheim to Millikan, telegram, July 12, 1926, Millikan, Box 16.6.

11 Guggenheim to Lee, July 14, 1926, Guggenheim Fund, Box 6.

12 Guggenheim to Millikan, telegram, July 14, 1926, Millikan, Box 16.6.

13 Guggenheim to Millikan, July 14, 1926, Millikan, Box 16.6.

14 Kármán to Millikan, telegram, July 26, 1926, Millikan, Box 16.6.

15 Kármán to Prussian Minister of Scholarship, Art, and Education ("Der Preussische Minister für Wissenschaft,

Kunst und Volksbildung"), hereafter "Ministry of Education," June 23, 1926, Kármán, Box 80.34.

16 J. de Kármán to M. Kármán, n.d., Kármán, Personal, Family File, Box 8. This and several other personal letters in Hungarian are also in English typescript, translated apparently in preparing *The Wind and Beyond*.

17 Kármán, *The Wind and Beyond*, p. 121.

18 Guggenheim to Millikan, July 19, 1926; W. F. Durand to Millikan, July 28, 1926; Millikan, Box 16.6. Millikan to Guggenhiem, July 13, 1926; Guggenheim to Durand, telegram, July 19, 1926; Guggenheim Fund, Box 6.

19 Millikan, copy of announcement in Pasadena, August 8, 1926, Guggenheim Fund, Box 6.

20 Kármán to Millikan, August 14, 1926; Kármán to Epstein, August 14, 1926; Kármán, Box 72.9. Kármán cites his telegram to Millikan in letter to Epstein.

21 Hallion, *Legacy of Flight*, p. 52.

22 Kármán, *The Wind and Beyond*, p. 122.

23 Ibid., pp. 124–26.

24 Richard P. Hallion, *Legacy of Flight*: The Guggenheim Contribution to American Aviation (Seattle: University of Washington Press, 1977), p. 52.

25 Kármán, *The Wind and Beyond*, pp. 128–29.

26 "Members of the educational conference called by the fund in Washington, D.C., on December 10 and 11, 1926," to the Daniel Guggenheim Fund, December 11, 1926, Kármán, Box 41.

27 Guggenheim to Kármán, December 9, 1926, Kármán, Box 11.40; also Kármán, Box 41.

28 Millikan to Guggenheim, November 5, 1926, Millikan, Box 16.6.

29 Kármán, *The Wind and Beyond*, p. 129.

30 Kármán wrote to H. I. Cone from Kobe on January 10, 1927, Kármán, Box 41.

31 Epstein wrote Kármán that he had heard of Kármán's "happy return," September 1, 1927, Kármán, Box 8.21.

CHAPTER 9
1 Prandtl to Kármán, April 23, 1920, Kármán, Box 23.

2 Paul A. Hanle, "Erwin Schrödinger's Statistical Mechanics, 1912–1925," Ph.D. dissertation, Yale University, 1975, pp. 166–68.

3 Mises to Prandtl, January 9, 1923, Mises. Mises noted the "well-known and much cited work written with Born on the specific heat of solid bodies." Also Erwin Schrödinger, "Spezifische Wärme (theoretischer Teil)," *Handbuch der Physik*, ed. H. Geiger and K Scheel, vol. X (Berlin: J. Springer, 1926), pp. 308–13.

4 Prandtl to Kármán, June 28, 1920, Kármán, Box 23.

5 Prandtl to Kármán, June 14, 1921, Kármán, Box 23.

6 Ibid.

7 Prandtl to Kármán, September 17, 1923, Kármán, Box 23.

8 Prandtl to Kármán, August 11, 1920, Kármán, Box 23.

9 Courant to Kármán, October 11, 1922, Kármán, Box 6.14.

10 Prandtl wrote Mises (January 3, 1923, Mises) that because he was going to Munich, the Aerodynamic Research Laboratory (Aerodynamische Versuchs-Anstalt, abbreviated AVA) would be separated from the chair and institute of applied mechanics, and that he had appointed his assistant Betz to direct the former.

11 Courant's remark is a cynical comparison to a group of whom German mathematicians bore singularly low opinions.

12 Karl-Heinz Manegold, "Universität, Technische Hochschule und Industrie: Ein Beitrag zur Emanzipation der Technik im 19. Jahrhundert unter besonderer Berücksichtigung der Bestrebungen Felix Kleins," *Schriften zur Wirtschafts- und Sozialgeschichte* 16 (Berlin: Duncker and Humblot, 1970); and Lewis Pyenson, "Einstein's Early Scientific Collaboration," *Historical Studies in the Physical Sciences* 7 (1976).

13 Max Born to Kármán, November 7, 1922, Kármán, Box 3.27.

14 Max Born, ed., *The Born–Einstein Letters* (New York: Walker, 1971), p. 73.

15 Theodore von Kármán and Lee Edson, *The Wind and Beyond* (Boston: Little, Brown, 1967), p. 69; a slightly different version in Kármán to Edson, November 4, 1961, Kármán, Box 115.6.

16 Ibid.

17 Courant to Kármán, January 30, 1923, Kármán, Box 6.14.

18 Prandtl solicited Mises' recommendation for the succession, which he could expect only to endorse Kármán, as it did in generous terms. Prandtl to Mises, January 3, 1923; and Mises to Prandtl, January 9, 1923; Mises.

19 Registrarial memoranda, July 28 and July 30, 1920, T. H. Munich, file "Technische Mechanik, Abschnitt a) v. Juli 1920–July 1921," and file "Technische Mechanik, Abschnitt b) v. Juli 1921–Oct. 1921."

20 Prandtl to M. Schröter, July 12, 1922, and Prandtl to Rektor of T. H. Munich, June 13, 1923 (in which he cites his previous acceptance of December 30, 1922), T. H. Munich, file "Technische Mechanik, Abschnitt c) v. Juli 1921–Dez. 1923."

21 Prandtl to Ministerium für Kultur und Unterricht (Bavaria), December 3, 1923, T. H. Munich, file "Technische Mechanik, Abschnitt c) v. Juli 1921–Dez. 1923."

22 "Denkschrift über die Einführung eines Lehrganges für Luftfahrtwesen an der Technischen Hochschule München," September 26, 1924, with cover letter from Akademische Fliegergruppe München e. V. to Bayerisches Staatsministerium für Unterricht und Kultus, September 27, 1924, Bayerisches Hauptstaatsarchiv, Bound files "Luftschiffahrt. Errichtung einer Reichsanstalt für Luftschiffahrt. Bildung einer bezw. Luftfahrt Controle," vol. I, 1910–1924.

23 Kármán, Box 23.

24 Kármán to Prandtl, December 7, 1923, Kármán, Box 23.

25 A. J. Ryder, *Twentieth-Century Germany: From Bismarck to Brandt* (New York: Columbia University Press, 1973), pp. 225–26.

26 Kurt Düwell, "Gründung und Entwicklung der Rheinisch-Westfälischen Technischen Hochschule Aachen bis zu ihrem Neuaufbau nach dem Zweiten Weltkrieg—Darstellung und Dokumente," in *Rheinisch-Westfälische Technische Hochschule Aachen 1870/1970*, ed. Hans Martin Klinkenberg, vol. I (Stuttgart: Oscar Bek Verlag, 1970), p. 93.

27 Klinkenberg, vol. I, p. 183; and Statistisches Landesamt of Prussia, *Statistisches Jahrbuch für den Freistaat Preussen* 23 (1927):174–75; ibid. 24 (1928):252–53.

28 *Statistisches Jahrbuch* p. 171.

29 Paul Forman, John L. Heilbron, and Spencer Weart, "Physics *circa* 1900: Personnel, Funding, and Productivity of the Academic Establishments," *Historical Studies in the Physical Sciences* 5 (1975): 61.

30 Gerhard Lüdtke, ed., *Minerva. Jahrbuch der gelehrten Welt* 28 (Berlin: W. de Gruyter, 1926):688–92.

31 Ibid., pp. 1–3.

32 Kármán, *The Wind and Beyond*, p. 136.

33 Kármán, memorandum draft to "Oberregierungsrat" of Saxon Ministry of Education, March 12, 1925, Kármán, Box 80.34.

34 Saxon Ministry of Education to Kármán, April 16, 1925, Kármán, Box 80.34.

35 Kármán's secretary to Saxon Ministry of Education, n.d., Kármán, Box 80.34.

36 Prussian Ministry to Kármán, May 25, 1925, Kármán, Box 80.34.

37 Kármán to "Oberregierungsrat Dr. Ulich," Saxon Ministry of Education, July 16, 1925, Kármán, Box 80.34.

38 Düwell, p. 97.

39 Ibid.

40 Ibid., p. 102; and carbon copy of internal memorandum of Ministry in Berlin, n.d., Kármán, Box 80.34.

41 Kármán to Prandtl, March 26, 1926, Kármán, Box 23; and on his consulting demands, Millikan to Guggenheim, telegram, July 9, 1929, Guggenheim Fund, Box 4.

42 Prandtl to Kármán, October 26, 1923, Kármán, Box 23.

43 Kármán to Prandtl, December 7, 1923, Kármán, Box 23.

44 Prandtl to Kármán, December 15, 1923, Kármán, Box 23. Prandtl wrote sarcastically, "One might conclude [from your response] that Aachen is a thousand miles further from Göttingen than is Delft."

45 Debye to Kármán, May 1, 1926, Kármán, Box 6.43.

46 Kármán to Debye, May 11, 1926, Kármán, Box 6.43.

47 Millikan to Guggenheim, December 6, 1928, Guggenheim Fund, Box 6.

48 Constance Reid, in *Courant in Göttingen and New York* (New York: Springer-Verlag, 1976), pp. 44, 125, has emphasized Kármán's position as a member of the "in group" of Göttingen mathematicians; a group, however, that itself had considerable difficulty after Klein's retirement in gaining acceptance into Göttingen academic society. Thus the rebuff of late 1922 suggests their own weak positions in faculty politics.

49 Paul Forman, "Weimar Culture, Causality, and Quantum Theory, 1918–1927: Adaptation by German Physicists and Mathematicians to a Hostile Intellectual Environment," *Historical Studies in the Physical Sciences* 3 (1971):4.

50 An indication is a remarkable letter to Kármán from J. M. Burgers in Delft, July 29, 1926, Kármán, Box 4.22, in which Burgers devotes, without introduction, the first six and one-half pages of the ten-page letter to a detailed mathematical exposition of Schrödinger's recent publications on wave mechanics.

51 Forman, "Weimar Culture, Causality, and Quantum Theory," pp. 46, 70–72, 80–82.

176 Notes

CHAPTER 10

1 Dekan der Fakultät für allgemeine Wissenschaften, Aachen, to Millikan, December 4, 1926, Kármán, Box 20.

2 Kármán to Paul Epstein, February 14, 1927, Kármán, Box 8.

3 Epstein to Kármán, September 1, 1927, Kármán, Box 8.21.

4 Kármán's secretary, Aachen, to Clark B. Millikan, March 20, 1929, Kármán, Box 20.

5 C. Millikan to Kármán, April 2, 1928, Kármán, Box 20.

6 Kármán to C. Millikan, August 5 and August 11, 1930, Kármán, Box 20.

7 Millikan to Kármán, January 24, 1927, Kármán, Box 20.

8 Millikan to Kármán, July 29 and August 26, 1927, Kármán, Box 20.

9 Millikan to Kármán, telegram, May 8, 1928, Kármán, Box 20.

10 Theodore von Kármán with Lee Edson, *The Wind and Beyond* (Boston: Little, Brown, 1967), p. 141.

11 Millikan to Guggenheim, December 6, 1928, Guggenheim Fund, Box 6.

12 Guggenheim to Millikan, telegram, December 7, 1928, Guggenheim Fund, Box 6. Evidently Millikan had discussed the issue earlier with Guggeinheim, for Millikan's letter could not have arrived in New York in one day.

13 Kármán, *The Wind and Beyond*, p. 141.

14 Millikan to Guggenheim, February 17, 1929, Guggenheim Fund, Box 6.

15 The "formal invitation to become Director" referred to in Millikan to Kármán, July 6, 1929, Guggenheim Fund, Box 4; and in Millikan to Kármán, August 26, 1929, Kármán, Box 72.9.

16 J. de Kármán to T. v. Kármán, n.d., Kármán, Personal, Family File, Box 8.

17 Kármán to Ministry of Education, Berlin, March 6, 1926, Kármán, Box 80.34, asking permission to live in Holland with sister and mother.

18 Kármán, *The Wind and Beyond*, p. 141.

19 Millikan to Kármán, July 6, 1929, Guggenheim Fund, Box 4; and Memorandum of Conference on Airship Institute, June 14, 1929, Guggenheim Fund, Box 4.

20 Memorandum of Conference, June 14, 1929, Guggenheim Fund, Box 4.

21 Millikan to Kármán, July 6, 1929, Guggenheim Fund, Box 4.

22 Memorandum of Conference, and Guggenheim, Hunsaker, and Millikan to Epstein, telegram, June 14, 1929, Guggenheim Fund, Box 4.

23 Millikan to Guggenheim, telegram, July 9, 1929, Guggenheim Fund, Box 4.

24 Memorandum of Conference.

25 *The World Almanac and Book of Facts*, 1933, "Foreign Exchange Rates, Yearly Averages," p. 412.

26 Guggenheim to Millikan, July 29, 1929, Guggenheim Fund, Box 4.

27 Guggenheim to Kármán, August 8, 1929, Kármán, Box 41.

28 Millikan to Kármán, August 26, 1929, Kármán, Box 72.9.

29 Guggenheim to Millikan, July 29, 1929, Guggenheim Fund, Box 4.

30 Epstein to Kármán, September 9, 1929, and October 6, 1929, Kármán, Box 8.21.

31 Kármán to Millikan, telegram draft, n.d., Kárman, Box 20.

32 Kármán to Millikan, October 9, 1929, Kármán, Boxes 20 and 41.

33 Guggenheim to Millikan, September 7, 1929, Millikan, Box 16.8.

34 Guggenheim to Millikan, October 11, 1929, Millikan, Box 16.8.

35 Millikan to Kármán, telegram, October 18, 1929, Kármán, Box 20; and Kármán to Millikan, telegram, October 20, 1929, Millikan, Box 16.8.

36 Düwell, p. 102.

37 Kármán to J. M. Burgers, December 12, 1929, Kármán, Box 4.22.

38 Kurt Düwell, "Gründung und Entwicklung der Rheinisch-Westfälischen Technischen Hochschule Aachen bis zu ihrem Neuaufbau nach dem Zweiten Weltkrieg—Darstellung und Dokumente," in Rheinisch-Westfälische Technische Hochschule Aachen 1870/1970, ed. Hans Martin Klinkenberg, vol. I (Stuttgart: Oscar Bek Verlag, 1970), p. 93.

39 Guggenheim to Millikan, October 22, 1929, Guggenheim Fund, Box 4.

40 Epstein to Kármán, April 5, 1930, Kármán, Box 8.21.

41 Kármán to J. M. Burgers, August 1, 1930, Kármán, Box 4.23.

42 Millikan to Kármán, June 11, 1930, Kármán, Box 20; and C. B. Millikan to R. A. Millikan, June 26, 1930, Millikan, Box 16.9.

43 Kármán, The Wind and Beyond, p. 146.

44 Prandtl to Kármán, April 24, 1931, Kármán, Box 23.

45 Kármán, The Wind and Beyond, pp. 142–44; and Düwell, p. 106.

46 Kármán to C. B. Millikan, August 11, 1930, Kármán, Box 20.

47 Prandtl to Kármán, September 29, 1932, Kármán, Box 23.

48 Düwell, p. 106.

49 Alan D. Beyerchen, Scientists under Hitler (New Haven: Yale University Press, 1977), pp. 1–39.

50 Ministry of Education to Rector of Aachen, May 4, 1933, copy in Millikan, Box 16.11.

51 Ministry of Education to Kármán, January 27, 1934, Kármán, Box 80.34.

52 Kármán, The Wind and Beyond, p. 146, my translation.

53 Prandtl to Kármán, July 7, 1933; and Kármán to Prandtl, August 2, 1933, Kármán, Box 23.

54 Kármán, *The Wind and Beyond*, p. 146; Kármán noted that Goering's office had suggested that he work for the air ministry as a consultant.

55 Ibid., p. 145.

56 Daniel J. Kevles, *The Physicists: The History of a Scientific Community in Modern America* (New York: Knopf, 1978); Charles Weiner, "A New Site for the Seminar: The Refugees and American Physics in the Thirties," in *The Intellectual Migration: Europe and America, 1930–1960*, eds. Donald Fleming and Bernard Bailyn (Cambridge, Mass.: Harvard Belknap, 1969), pp. 190–234; Paul Forman, John L. Heilbron, and Spencer Weart, "Physics *circa* 1900: Personnel, Funding, and Productivity of the Academic Establishments," *Historical Studies in the Physical Sciences* 5 (1975).

57 Kármán wrote Epstein on February 3, 1930, Kárman, Box 8, observing only partly in jest, "It is said here that half of America has collapsed. I only hope that you and our common friends belong to the other half of the Americans."

58 See Chapter 4 and notes 18 and 19. Also, support for Kármán's dissertation is cited in "Bericht des Herrn Professor Dr. Prandtl betr. Institut für angewandte Mechanik," part of financial report of Göttinger Vereinigung for 1909, Nachlass Klein, file 4F "Göttinger Vereinigung 1908–1912."

59 Kármán, *The Wind and Beyond*, pp. 85, 222.

60 Ibid., p. 272.

61 Ibid., as suggested by account on p. 268.

62 Ibid., p. 352. "For me as a scientist the military has been the most comfortable group to deal with, and at present I have found it to be the one organization in this imperfect world that has the funds and spirit to advance science rapidly and successfully."

63 Nathan Reingold, "The Case of the Disappearing Laboratory," *American Quarterly* (Spring 1977):101.

CHAPTER 11
1 Leo Szilard, John von Neumann, Edward Teller, Eugene Wigner, Cornelius Lanczos, et al. See William O. McCagg, *Jewish Nobles and Geniuses in Modern Hungary* (New York: Columbia University Press, 1973).

2 *The Guggenheim Aeronautical Laboratory of the California Institute of Technology: The First Twenty-Five Years* (Pasadena: California Institute of Technology, 1954), p. 7.

INDEX

Aachen, Aerodynamics Institute at,
54, 66–67
and expansion issue, 115–120, 130
and Peace of Aachen, 29, 38
Abraham, Max, 30
Ackeret, Jakob, 51
Aerodynamical Laboratory (Paris), 13
Aerodynamics
and airplane design, 91–92
American growth in, 152–157
as applied science at Göttingen,
36–41
boundary-layer hypothesis in, 43–48,
58, 60, 90
Caltech grant for, 11–22
German accomplishments in, American recognition of, 81–90
German work summarized (1913–1926), 62–64, 107
Munk's contributions to, 85, 90–91
studies, American vs European,
13–14, 80–81, 84
vortex street theory in, 54–61
wing theory in, 49–51
Aerodynamic Theory, 40
Aeronautical Engineering, 85
Aeronautics
engineering/industry alliance in, 4–8
and foundations, 10, 13
at Göttingen, 36–41, 81
Air Commerce Act, 2, 84
Althoff, Friedrich, 31, 37
Ames, Joseph S., 86–90, 92, 93, 155
Anti-Semitism, 109–112, 121–122,
132–133

"Applications of Modern Hydrodynamics to Aeronautics"
(Prandtl), 86
Applied mechanics, 24–26
Applied science
at Göttingen, 22, 24–36, 51–52
and Kármán's research methods,
65–67
and Prandtl's research methods,
42–44, 47–49
Arnold, Henry H., 136, 137
Arnstein, Karl, 129, 135
Atlantic Aircraft (Fokker), 3, 136
Aviation, 85–86
Aviation industry
and foundations, 10
overview of, 1918–1926, 1–4
and universities, 4–8

Bairstow, L., 89, 90
Bateman, Harry, 14, 15, 21, 94, 96, 98
Beech Co., 3
Bell, Edward T., 14, 21
Betz, Albert, 51, 109
Bienen, Theodore, 63
Bliss, Colins, 6
Boeing, William, 4
Born, Max, 30, 40, 99, 105, 106,
108–113, 122
Böttinger, Henry Theodor, 31–32,
38, 39
Boundary-layer hypothesis, 43–48,
58, 60, 90
Brown, Elmer E., 4
Burgers, Jan, 131